Tim Peake's view from the International Space Station. Looking to the east along the English channel with the UK on the left and France on the right.

TIM PEAKE
& THE EUROPEAN SPACE AGENCY

THE
ASTRONAUT
SELECTION
TEST BOOK

with Colin Stuart

Illustrations by Ed Grace

CENTURY

1 3 5 7 9 10 8 6 4 2

Century
20 Vauxhall Bridge Road
London SW1V 2SA

Century is part of the Penguin Random House group of companies whose addresses can
be found at global.penguinrandomhouse.com.

ILLUSTRATIONS BY ED GRACE - WWW.EDGRACE.CO.UK

First published by Century in 2018

www.penguin.co.uk

A CIP catalogue record for this book is available from the British Library.

ISBN 9781780899183 (hardback)
ISBN 9781780899190 (trade paperback)

Printed and bound in Great Britain by Clays Ltd, Elcograf S.p.A.

Penguin Random House is committed to a sustainable future for
our business, our readers and our planet. This book is made from
Forest Stewardship Council® certified paper.

To the first person to set foot on Mars. Godspeed!

To all of those on Earth who make human space exploration possible.

Thank you.

CONTENTS

'You need to live a little bit outside your comfort zone because you can be even more than you dream of ... I figure if a farmer's daughter from Iowa can become an astronaut, you can be just about anything you want to be.'

PEGGY WHITSON

With a total of 665 days in space, Peggy Whitson holds the US record for the most cumulative time in space. In her career as a NASA astronaut she performed ten spacewalks and was the first female commander of the International Space Station.

INITIALISING TEST PROCEDURE

TEST 1

Astronaut Candidate, you have crash-landed on the Moon.

You and your two crew members were originally scheduled to rendezvous with a lunar base. However, due to mechanical difficulties, your spacecraft was forced to land at a spot some 50 km off course. During landing, the spacecraft and much of the equipment aboard was damaged and, since your survival depends on reaching the lunar base, the most critical items must now be chosen for the trip. The good news is that in lunar gravity you can cover the terrain at about 5 km/h. The bad news is that your spacesuit consumables will only last eight hours.

Below you will find a list of the 15 items left intact and undamaged after landing. Your task is to rank them in terms of their importance in enabling your crew to reach the lunar base. Place the number 1 by the most important item, the number 2 by the second most important, and so on, through to number 15 for the least important.

You have two minutes. Godspeed!

- Box of matches
- Handheld GPS receiver
- 15 metres of nylon rope
- Three spare batteries for spacesuit
- Parachute silk
- Three spare carbon-dioxide removal canisters for spacesuit
- One case of dehydrated food
- Three spare oxygen tanks for spacesuit
- Stellar map
- Self-inflating life-raft
- Magnetic compass
- 20 litres of water – can be administered through special drinking port in spacesuit
- Two handheld mirrors
- First-aid kit, including medical tape, scissors, etc.
- Solar-powered FM receiver-transmitter

TEST 2

Thanks to expert prioritisation, your team of astronauts arrived safely at the lunar base. The next day your objective is to pilot a lunar exploration vehicle to a site of geological interest to excavate precious rock samples. It would normally take nine hours to reach your destination at an average speed of 72 km/h. But you have limited daylight and you must get there sooner. How long would the vehicle take to arrive at its destination if you travelled 8 km/h faster?

You have one minute to calculate the correct answer. You cannot use a calculator. The four individuals in your team have made their own estimates, as outlined in answers a–d below. Which answer is correct?

a) 8 hours and 48 minutes
b) 7 hours and 50 minutes
c) 8 hours and 6 minutes
d) 8 hours and 53 minutes

TEST 3

Well done – your team reached the geological Moon site in time. Unfortunately the crane on the back of your lunar exploration vehicle, used to lift the rocks, is damaged, so you need to design a new pulley system.

1. Which construction is more useful for lifting 400 kg?

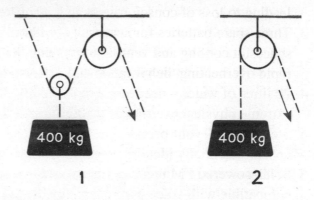

a) 1
b) 2
c) There's no difference

2. In another attempt to fix the crane, Mission Control is suggesting that you devise a clockwork motor. They have sent you a diagram of the system (below). Wheels 1 and 2 have the same radius. How fast will Wheel 2 rotate, if Wheel 1 is being driven?

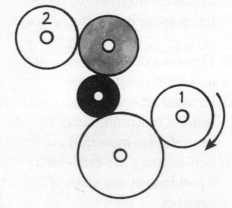

a) As fast as Wheel 1
b) Faster than Wheel 1
c) Slower than Wheel 1
d) Wheel 2 does not rotate

Answers

Test 1

The correct ranking is:

1. Three spare oxygen tanks for spacesuit – required not just for breathing, but also to maintain pressure in the suit.
2. Three spare carbon-dioxide removal canisters for spacesuit – without these, you will suffer carbon-dioxide poisoning, leading to loss of consciousness and, ultimately, death.
3. Three spare batteries for spacesuit – without power, the spacesuit cooling and ventilation systems will fail, leading to rapid overheating, dehydration and exhaustion.
4. 20 litres of water – needed for replacement of fluid, due to extreme physical exertion of the journey.
5. Stellar map – your primary means of navigation; star patterns appear essentially identical on the Moon as they do on Earth.
6. Solar-powered FM receiver-transmitter – even if not compatible with voice communication from inside a spacesuit, it could be used to communicate with a rescue party via Morse code, using the carrier signal; limited to short-range transmissions only.
7. Two handheld mirrors – a secondary method of signalling and communication, particularly if the group has to split up.
8. Parachute silk – useful for makeshift constructions (e.g. a hammock stretcher for an incapacitated crew member, a sack for carrying items, etc.) or as added protection from the Sun's rays.
9. 15 metres of nylon rope – useful for scaling obstacles and for makeshift constructions.
10. First-aid kit – while most medication cannot be administered inside a spacesuit, items such as tape, scissors, and so on, can be useful for makeshift constructions.
11. Self-inflating life-raft – bulky to carry (but weight is not such a problem, in one-sixth of Earth's gravity); could function as a stretcher.

The following items could be ranked in any order – none of them would be of much use for this task.

12. Handheld GPS receiver – on the moon this would not receive the weak signal from GPS satellites and, even if it did, the distance is so great that the triangulation method used to determine location would result in an accuracy of only tens of kilometres.
13. Magnetic compass – the magnetic field on the Moon is not polarised, so the compass will not work and will not aid your navigation.
14. One case of dehydrated food – there is no means to administer dehydrated food into the spacesuit.
15. Box of matches – virtually worthless; there is no oxygen on the Moon to sustain combustion.

Test 2

c) 8 hours and 6 minutes.

Test 3

1. a) Construction 1 is better. This type of double pulley is called a 'Gun Tackle', and will halve the amount of force you need to lift the weight.
2. a) As fast as Wheel 1.

TEST DEBRIEF

Congratulations, you have just completed the first test on your way to becoming an astronaut. The questions above are all based on real tests that astronauts are given as part of the rigorous Astronaut Selection Test recruitment for the European Space Agency (ESA).

How did you perform in the first test? Would your team have survived? The scenario is designed to test a candidate's situational awareness and decision-making aptitude – both crucial skills for an astronaut to demonstrate in the dynamic environment of space. Your order of ranking does not have to be exact, as long as you applied good logic, critical thinking and timely decision-making.

During the second test, did you notice that the time pressure added to the difficulty of the problem? This question tests an astronaut's mental arithmetic and mathematics skills. Astronauts are sometimes called upon to make split-second calculations to adjust for external variables or changes to mission plans. Even if an astronaut does not know the exact number or answer, he or she will try to make a quick estimation. A best guess in a pressurised situation may just make the difference between surviving and not surviving in space.

Did you get both of the correct answers in the third test? An engineering and visualisation aptitude is vital for any astronaut, as missions often require astronauts to operate or repair equipment they can't fully see, and they must do so in microgravity, which affects their senses, as there is no fixed up or down in weightlessness.

These are not the only skills that an astronaut must possess in order to get a coveted mission to space. To pass astronaut selection and training, candidates must demonstrate leadership abilities, excellent teamwork, memory retention, concentration, visual perception, foreign-language aptitude (in particular Russian), operational skills, technical prowess and physical health and fitness, to name just a few of the qualities. In short, an astronaut must be a jack of all trades: pilot and passenger; doctor and patient; scientist and subject; explorer and follower.

INTRODUCTION

The book in your hands is the manual that I wish I had been given when I applied to be an astronaut.

Being an astronaut is a job like no other. **Of the estimated 100 billion people who have ever lived, only 557 people have travelled to space (at date of writing)**. While it often features highly on lists of dream careers, there aren't many occupations with fewer available places. I applied with thousands of other people, but only six of us were successful in 2009. That was the first European astronaut selection to take place in a decade. You can't simply go to university to study how to become an astronaut. A careers advisor would be hard pressed to recommend a guaranteed career route – there isn't one.

But while becoming an astronaut is not easy – indeed it has been the hardest single thing I've ever done – there's also never been a better time to consider becoming a space farer. We are on the cusp of incredibly exciting missions to the Moon, Mars and beyond. Our technology is improving at a rapid pace, and so too is the science about long-duration human spaceflight. Thanks to international space agencies and private aerospace companies working closely together, there may soon be more seats to space than ever before.

The aim of this book is to demystify the astronaut selection and training process, and to identify and test candidates on what it takes to be an astronaut today. It follows closely the stages and criteria of the last ESA Astronaut Selection Test that was conducted, as well as the training programmes that ensued.

Part quiz, part guide, this book includes real puzzles, tests and exercises that ESA astronauts are given, ahead of missions to the International Space Station (ISS). It also explores what is likely to be involved in future astronaut selections for longer-duration missions to the Moon, Mars and beyond. The book is structured in four sections, and spans the stages of the ESA Selection Test:

- Round 1: 'Hard skills'
- Round 2: 'Soft skills'
- Round 3: Medical tests
- Round 4: Job interview.

In Part One, you will have to overcome one of the most discerning stages of the selection process to become an astronaut. You will be tested on a wide range of 'hard' skills for which astronauts must show a strong aptitude. Hard skills are specific (often non-trainable) abilities that can be measured easily – although you can make some improvement in performance if you practise; there are tests for visual perception, visual memory, maths, concentration and some psychological profiles.

Part Two looks at the other stages of the astronaut selection process. Here you will find out about the rigorous requirements for becoming an astronaut – physical, psychological and more general. You can answer the questions on the 2008 ESA Astronaut Selection application form, and find out what ESA is looking for in potential candidates.

In Part Three – once you have been selected – you will embark on astronaut training. This is a process that normally takes several years and spans subjects related to human behaviour and performance. You will be tested on your 'soft' skills. These are broadly your own personal attributes that enable you to interact well with others. You will be tasked with teamwork exercises, survival training, spacewalk (extravehicular activity or EVA) instruction, communication and language skills, and with preparation for living and working in microgravity.

In Part Four, we will look to the future and will explore the unique challenges and requirements for astronauts on long-duration missions to the Moon and Mars – and even further afield. We will examine some of the fascinating isolation experiments that have been conducted on Earth to simulate future human spaceflights, such as Mars-500 (see page 224), when six crew members lived in a mock-up spacecraft for 520 days, with extremely limited communication with the wider world.

There are also two photo inset sections in the book. The first depicts survival training and CAVES, initiatives which form part of an astronaut's preparation for space. The second photo inset section examines advanced mission training, including Neutral Buoyancy Training, NEEMO, Centrifuge Training and Zero-g Training.

If you are an aspiring astronaut, I hope you will find the book useful, challenging and entertaining. Equally, if you are perfectly content to live on our beautiful planet Earth, I hope you will take some enjoyment from the taxing tests and procedures.

Every human spaceflight mission requires the support of an incredible network of ground crew at Mission Control. This book is no different, and is a true collaboration. I have jointly written the book with a wide cast of ESA astronaut trainers and former ESA astronauts. In no particular order, the contributors to this book include: psychologists, spacewalk instructors, astronaut selection and training coordinators, flight directors, medical doctors, leadership and team-building experts, survivalists, communications experts, robotics and spacecraft piloting instructors, centrifuge operators, linguists, IT and operational managers, scientists and aeronautical engineers.

The full team, without whose utter commitment this book would not have been possible, is listed and thanked in the Acknowledgements.

HOW TO USE THIS BOOK

Each section of the book includes a range of astronaut tests, which vary in difficulty. Answers to the puzzles and questions follow each test, along with a score rating system, which indicates how your results stack up against the results of a real astronaut. More subjective, open-ended answers are discussed here too.

Note: Please read each question carefully before you try to answer it. Astronauts must demonstrate excellent attention to detail. Mistakes are often made not through a failure of intelligence, but through a failure to grasp what is actually being asked.

Each section of the book also includes wider background information on becoming an astronaut, and boxes labelled 'Astronaut essentials' in Part Three emphasise pertinent skills and techniques.

All that's left for me to say now is good luck! Do you have what it takes for space? Find out in T-minus 5...4...3...2...1!

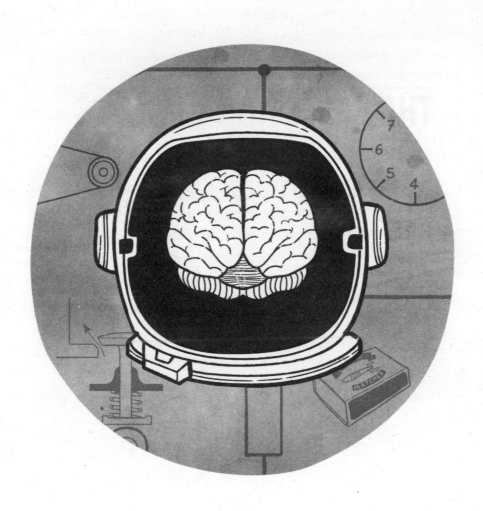

PART ONE
THE SELECTION PROCESS

ROUND 1: HARD SKILLS

'The test programme puts high demands on your ability to keep up a high level of attention and concentration and your physical and psychological stress resistance. Therefore, it is of paramount importance that you are relaxed and healthy. We recommend not using any medication, since these may have an adverse effect on your cognitive functions.'

ESA SELECTION TEST ADVICE TO ASTRONAUT
CANDIDATES IN 2008

Astronaut candidate, the initial round of 'hard skill' testing focuses on your cognitive capabilities and psychomotor performance. 'We are looking for how your brain is hard-wired,' says former astronaut Gerhard Thiele, who was involved in the 2008 ESA astronaut selection process. 'Can you digest huge amounts of information? Can you select the critical information from the non-relevant? And how quickly do you process these kinds of things?'

The tests in this round are divided into the following skill-group sets: spatial awareness, visual perception, memory retention, technical information, concentration skills, English-language skills, mental arithmetic and measurement exercises. Try them for yourself.

You will notice that there are some time limits for these questions, and these simulate ESA test conditions. In 2008, Astronaut Selection Test candidates were asked to answer the questions and puzzles in this round as part of a long day, comprising six sessions of computer-based tests. The schedule for the day was as follows:

08.00–08.30	Registration
08.30–09.00	Introduction
09.00–09.45	Tests I
09.45–09.55	Break
09.55–11.00	Tests II
11.00–11.10	Break
11.10–12.10	Tests III
12.10–13.10	Lunch break
13.10–14.00	Tests IV
14.00–14.10	Break
14.10–15.25	Tests V
15.25–15.35	Break
15.35–17.00	Tests VI
17.00–18.00	Closing

Although this book can't exactly replicate the testing day shown above, try keeping to the recommended time limits when given.

ADVICE FOR RECRUITS

The first round demands intense concentration. I recommend taking short breaks between the different types of questions in this round, although – as on my day of testing in 2008 – such breaks may be too brief to feel fully rested or prepared for the next evaluation. This mental exertion is part of the test.

It is important not to dwell on your performance. If you think you have done badly in one test, you must put it out of your mind and simply focus on the next task in hand. Under these circumstances, a relaxed approach is the best way forward, but that of course is a skill in itself – being able to relax under intense pressure when the stakes are high.

In 2008 I was sitting in a room with many of my fellow candidates to take the test. If you're not careful, you can quickly become distracted by others, particularly if you see them getting through the questions faster than you are. I very much saw the whole process as a competition against myself, rather than against anyone else, and so I focused solely on what I was doing. That said, I'd be lying if I said I wasn't intimidated by sitting in a room with people who had doctorates in science, medicine and engineering, when I left school at 18 and joined the army! However, the tests are carefully designed to evaluate intellectual capacity, regardless of academic background. There are many ways to develop these kinds of skills and, as it turned out in my own case, years of operational experience of flying helicopters had stood me in good stead. Similarly, for this round, you do not need to have experience of one particular discipline to attempt the following 'hard skill' tests.

If you do attempt these questions alongside a friend, family member or colleague, it will better re-create the sense of competition that is part of the real Astronaut Selection Test. But ultimately you are competing against yourself, to bring out your optimum performance, so completing the questions alone is just as pertinent.

Note: Some questions have additional time limits, while others do not. This is to reflect the real ESA test. It leads to an interesting decision for candidates to make: are the selectors looking for speed or accuracy? Do you finish all the exercises, perhaps sacrificing getting them all correct? Or do you ensure that you get as many as possible correct, and disregard the number that you finish?

I won't spoil things by giving you the answer now, but bear all these factors in mind as you tackle the questions. On page 60 I'll give you a debrief, so that you can see how your performance measures up.

Good luck!

SPATIAL AWARENESS TEST

Spatial awareness skills are enormously important for any astronaut, because astronauts often operate equipment they can't actually see, and are doing so in microgravity, which affects their perception (after all, there is no fixed up or down in weightlessness). On a spacewalk in particular, you are constantly changing your orientation and perspective as you move around outside the International Space Station. It's vital to be able to build mental models of what you are doing, where you are going and how best to perform a task. It's actually a very liberating feeling, once your brain accepts that it can choose any orientation to work in. Selectors are looking for good spatial awareness skills right off the bat.

TEST 1

Imagine that you are facing a cube. This cube can roll to the left, right, forward (towards you) or backwards (away from you). There is a dot on the bottom of the cube.

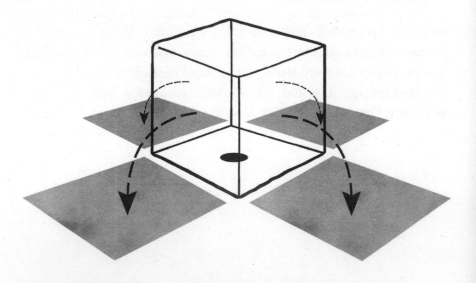

a) In your mind, roll the cube: forward, left, left, forward, right, backwards, right. Where's the dot now?

b) Imagine the same cube with the dot. Roll the cube: forward, right, right, forward, left, backwards, left. Where's the dot now?

Score 1 point for each correct answer.

Hint: During the real spatial awareness test, astronaut candidates hear someone read out the directions of rotation, and successfully navigating a puzzle like this not only requires you to come up with a strategy, but to be willing and able to adapt that strategy. For example, you might start by rotating the cube in your mind's eye after you hear each new instruction. However, if halfway through the test the speed at which the instructions are read quickens – too fast to rotate with each new word – then you need to adapt and find a new method. Perhaps you're able to remember the remaining instructions and then rotate the cube afterwards. Astronaut selectors are always looking for adaptable people who are calm under pressure.

VISUAL PERCEPTION

Modern spacecraft and the ISS are full of instrument panels, dials, monitors and displays. It's important for astronauts to be able to assimilate information quickly and recall data with speed and accuracy. One day, this could save you and your crew.

TEST 2

Below are a series of instrument dials. You are allowed to look at each set of dials, which includes nine readings, for just three seconds. Your task is to identify and read the numerical values of the 'critical' instruments in a specified order – from left to right, beginning with the first line. The instruments show different readings and are different in a prominent feature, such as shape or colour. Prior to each presentation you are informed what constitutes a critical instrument. For example, 'Critical: round, white instruments'.

Look just at the nine dials in question, and cover all other sets with a piece of paper. Then move the paper to cover all the dials. Remember, you only have three seconds! In the real ESA test these dials will flash on a computer screen for three seconds, but testing a candidate's honesty in this version of the test is also instructive – you are only cheating yourself if you take longer.

Here is a sample test before you begin. 'Critical: black instruments':

Answer: 2, 8, 2, 5.

a) Critical: white instruments (1 point for each you can remember):

b) Critical: round, white instruments (1 point for each you can remember):

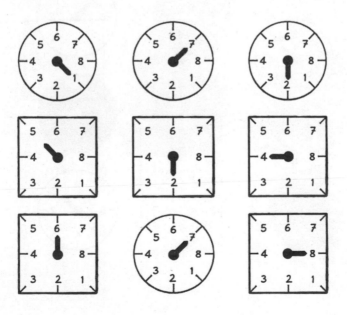

c) Critical: round, black instruments (1 point for each you can remember):

d) Critical: square, white instruments (1 point for each you can remember):

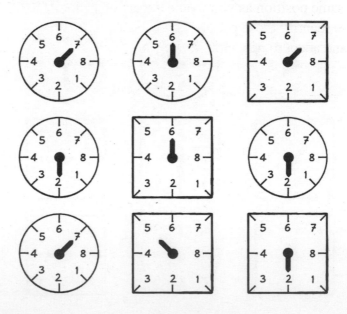

e) Critical: black instruments (1 point for each you can remember):

Hint: Did you develop a good technique for scanning only the critical instruments? Were you distracted by the numbers not being in the same position as you would expect on a clock face? This test helps to identify those people who can easily discard preconceived ideas and adapt to new situations.

MEMORY RETENTION

Astronauts are often asked to remember and relay sequences of numbers – pressures, temperatures, coordinates, etc. – to Mission Control. To test their memory skills before they are selected, they are read a long sequence of numbers. When the voice reading the numbers stops, they must repeat the sequence of numbers, backwards, trying to recite as many numbers as they can recall. The problem is that you never know when the voice is going to stop! To make this task even more demanding, the voice reading the numbers varies in rhythm and pitch, so it is very hard to develop a strategy for remembering the sequence.

TEST 3

Ask someone to read to you the numerical sequences below, and try reciting the numbers backwards.

How many digits you can remember? You cannot use a pen and paper, and you have ten seconds for each answer. You get 1 point for each digit that you can successfully recall.

a) 8 6 4   f) 9 2 5 7 4 7 3 4

b) 7 2 3 5   g) 2 6 7 1 0 2 8 4 6

c) 9 3 0 2 1   h) 4 3 7 8 1 2 9 6 5 5

d) 0 7 4 3 5 7   i) 2 3 5 7 9 4 8 6 1 2 0

e) 2 7 3 1 9 0 4   j) 6 1 5 8 0 4 2 4 6 2 6 4

Hint: Were you able to find some way of grouping sets of numbers together in your mind, to make it easier to remember them? If you found that too easy, try doing it again while you step up and down the first step of a staircase. Astronauts often have to exercise good memory retention while simultaneously doing a physically demanding task. Spacewalking is a good example of this – and of course there's no pen and paper to hand on a spacewalk to write things down.

Note that most adults, without training, can remember and repeat backwards sequences of numbers six digits long. Some astronauts can remember backwards sequences upwards of 12 digits long.

VISUAL MEMORY

Being able to remember something straight after you've heard it is one skill, but it also pays to remember it long after you've seen it.

TEST 4

Have a look at the following numbers. Each number is accompanied by a shape. You have a maximum of three minutes to remember these number/shape combinations. Even if you cannot remember them all, try to remember some of them. You will be tested on them later in this section.

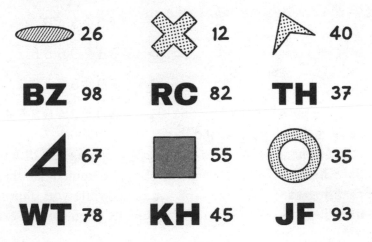

TECHNICAL INFORMATION

An astronaut has to be a bit of everything: a doctor, an engineer, a pilot, a leader. While your training will equip you with all the technical skills you need in order to survive in space, you're expected to have a basic understanding of some scientific and engineering principles. A large part of an astronaut's life on board the International Space Station is spent maintaining equipment. After all, the ISS has been in space now for more than 20 years and it has to endure punishing conditions: large thermal extremes between day and night; harsh exposure to radiation; and occasional strikes from micrometeorites and pieces of space debris. Astronauts frequently have to maintain electrical systems, thermal cooling systems and computer technology. And, of course, when the loo malfunctions, you'll need to be the plumber, too!

The tests below will give you a good idea of your knowledge of basic electrical circuits, engines, physics and tools. Have a go at these questions and see how you get on. You get 1 point for each correct answer.

TEST 5

1. Which part does not let direct current pass through?
   a) Resistor
   b) Coil
   c) Copper cable
   d) Capacitor

2. What is the speed of sound in dry air at 0°C?
   a) 150 m/s
   b) 330 m/s
   c) 240 m/s
   d) 3500 m/s

3. Which material does a lightning conductor usually consist of?
   a) Plastic
   b) Copper
   c) Lead
   d) Tin

4. What is the voltage in a European electrical socket?
   a) 180 volts
   b) 60 volts
   c) 50 volts
   d) 220/230 volts

5. What is the reading on the voltmeter?

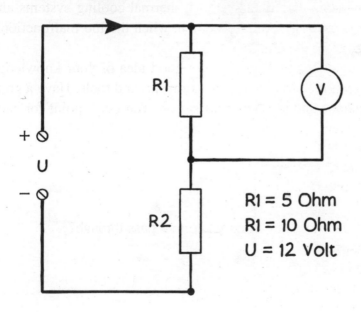

R1 = 5 Ohm
R1 = 10 Ohm
U = 12 Volt

   a) 2 volts
   b) 4 volts
   c) 4.5 volts
   d) 6 volts

6. Which path will the ball take if it is cut loose at the indicated place?

a) None
b) 1
c) 2
d) 3

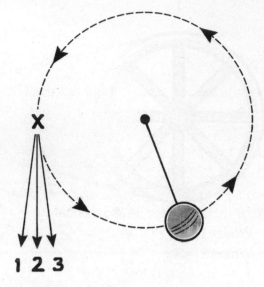

1 2 3

7. Which of the three circuits are closed?

a) A and B
b) A and C
c) A, B and C
d) A

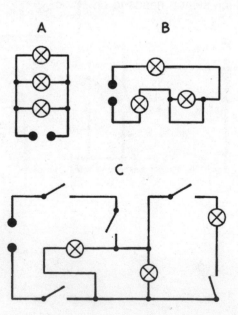

A

B

C

8. Which of the two wheels rotates faster?

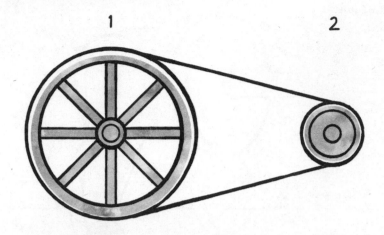

1

2

a) Wheel 1
b) Wheel 2
c) Both are equally fast
d) It depends on whether the propulsion is attached to Wheel 1
   or Wheel 2

9. Which switch needs to be closed to allow current to flow?

a) A
b) B
c) C
d) D

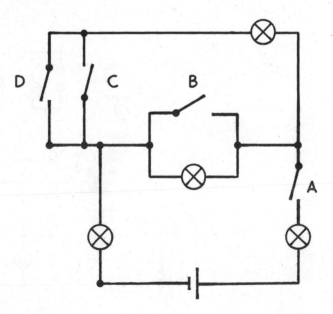

10. In which direction does Axis 2 turn when Axis 1 is turned as indicated?

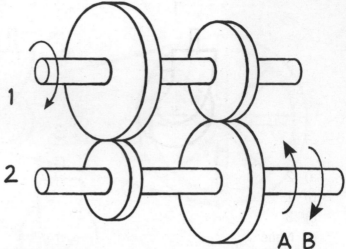

a) A
b) Axis 2 doesn't turn
c) Small wheel A, large wheel B
d) B

11. Across which of the wheels does the drive belt have to be tightened for the top axis X to rotate fastest?

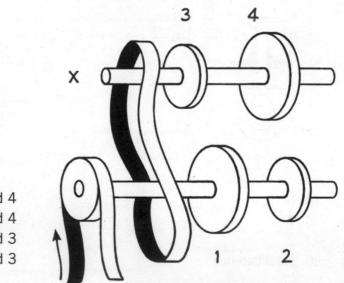

a) 2 and 4
b) 1 and 4
c) 1 and 3
d) 2 and 3

12. Which is the most useful swivel-chair wheel?

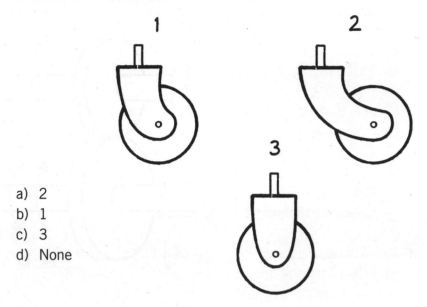

a) 2
b) 1
c) 3
d) None

13. Which valve lets through the greatest gas volume during one cam rotation?

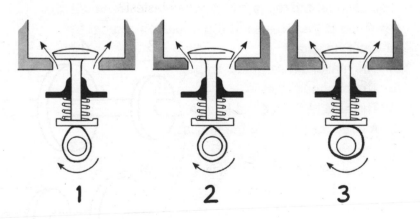

a) 3
b) 1
c) 2
d) No difference

14. Name the four power strokes of an Otto four-stroke engine
    a) intake – compression – exhaust – ignition
    b) compression – ignition – intake – exhaust
    c) intake – compression – ignition – exhaust
    d) intake – ignition – compression – exhaust

15. What is the engine oil needed for, apart from greasing the movable parts of an engine?
    a) Reduction of the engine temperature
    b) Keeping the engine temperature constant
    c) Balancing the mass
    d) Balancing the weight of the vehicle

16. Approximately how much engine oil is in a mid-size car?
    a) 0.5–1 litre
    b) 1–2 litres
    c) 4–6 litres
    d) 10–15 litres

17. What is the significance of the red line representing the beginning of the red area of the revolution counter (rpm indicator) in a car?
    a) The maximum permitted rpm
    b) The minimum rpm
    c) The area of the highest torque
    d) A visual mark to find the indicator

18. What fuel are aircraft usually filled up with?
    a) Diesel
    b) Petrol
    c) Octane
    d) Kerosene

19. What is the pressure of an average motor-car tyre?
    a) 0.1–0.8 bar
    b) 1.8–2.8 bar
    c) 3.5–7 bar
    d) 24–30 bar

20. Which temperature unit does not exist?
    a) Calvin
    b) Kelvin
    c) Celsius
    d) Fahrenheit

21. In a circuit the voltage, U, is increased evenly. The resistance of 100 ohms stays constant. What happens to the current, I?
    a) Increases with voltage
    b) Decreases with voltage
    c) Increases at a rate of 75 per cent of the voltage
    d) Does not change

22. In a circuit the resistance, R, is proportionally increased. The voltage (U) of 5 volts is constant. What happens to the current, I?
    a) It increases by R
    b) It increases by $R \times 1.5$
    c) It is reduced
    d) It increases by the square of R

23. Which statement does not apply? The electrical voltage, U . . .
    a) is the pressure on free electrons
    b) is the cause of electrical current
    c) is the result of the effort to balance electrical charges
    d) can be measured in Fahrenheit

24. Which kinds of current can be distinguished?
    a) Direct and alternating
    b) Direct, alternating and quick
    c) Direct and solar
    d) Alternating and quick

25. Which is the right measure for the rate of flow of electrons through a circuit (I)?
    a) Joule
    b) Ampere
    c) Watt
    d) Fahrenheit

26. Which formula is the correct one for calculating the voltage?
    a) Resistance minus current
    b) Resistance divided by current
    c) Resistance plus current
    d) Resistance multiplied by current

27. Which (electrical) condition is characteristic of a short circuit?
    a) The positive and negative poles of a voltage source have a direct connection (= 0.8 ohm)
    b) The positive and negative poles of a voltage source have a direct connection (= 10 ohm)
    c) The positive and negative poles of a voltage source have a direct connection (= 0 ohm)
    d) The positive and negative poles of a voltage source have a direct connection (= 1 ohm)

28. Which device is best suited to support the gate?

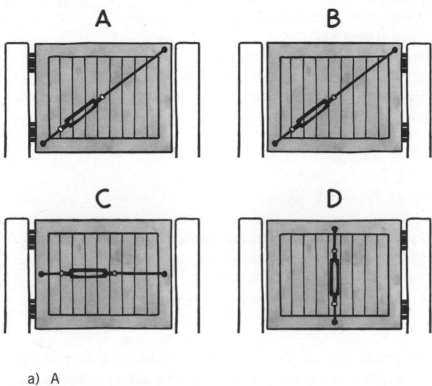

a) A
b) B
c) C
d) D

CONCENTRATION SKILLS

In space it is vital for astronauts to have supreme concentration skills. Tasks are often long and repetitive, as well as requiring you to process key technical information. Spacewalks, for example, can last more than eight hours. You don't want your concentration levels dropping, so attention to detail is paramount.

TEST 6

In this test of your powers of concentration you are presented with a series of triangles. Each triangle has three traits to consider, with different possible options for each trait:

- **Orientation** – the tip of the triangle can be pointing up, down, left or right.
- **Dots** – the triangle can contain zero, one, two, three or four dots.
- **Shading** – the triangle can be none, spotted, solid or striped.

Here are some sample triangles:

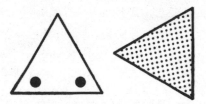  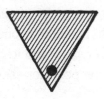

You will then be shown some working rules that you have to memorise in a limited amount of time. For example, you could be told:

- If two triangles with the same number of dots succeed each other, tick the box above the triangle (see A1–A2 below).
- If two triangles of the same shade succeed each other, tick the box below the triangle (see B1–B2 below).
- If neither rule applies, tick the box to the side of the triangle (see C1–C2 below).

Examples:

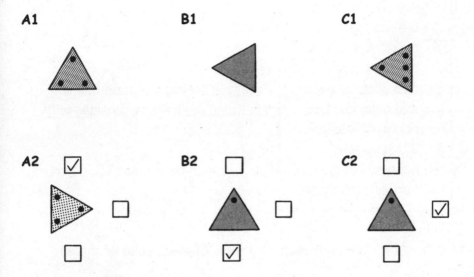

In the following questions, cover over subsequent triangles with a piece of A4 paper with a square cut into the middle, large enough to reveal only one triangle and the boxes around it. Work down the columns, top to bottom, revealing the triangles one by one. Score one point for each box ticked correctly.

Question 1

Rules (you have ten seconds to memorise these)

- If two triangles with the same position succeed each other, tick the box above the triangle.
- If two triangles with the same number of dots succeed each other, tick the box below the triangle.
- If neither rule applies, tick the box next to the triangle.

2

3

4

5

1

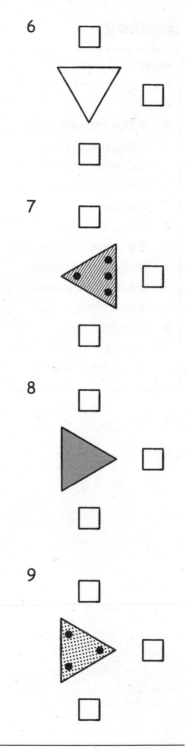

Question 2

Rules (you have ten seconds to memorise these)

- If two triangles of the same shade succeed each other, tick the box above the triangle.
- If two triangles of the same position succeed each other, tick the box below the triangle.
- If neither rule applies, tick the box next to the triangle.

3

4

1

5

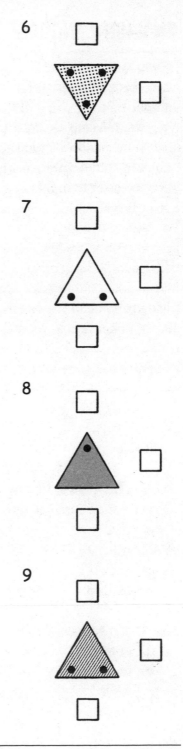

ENGLISH-LANGUAGE SKILLS

The International Space Station has two official working languages: English and Russian. As part of your training you will learn the latter, but before you are selected as an astronaut you have to prove that your existing English-language skills are up to the required standard. If you are a native English speaker, you may find this section easier than other sections of the book. But remember, ESA astronauts from its 22 Member States must all pass this test.

TEST 7

Here are some of the questions that were used to put applicants through their paces. You get 1 point for each correct answer.

1. Peter has gone to the station, . . . he?
   a) hasn't
   b) isn't
   c) didn't
   d) doesn't

2. Find a synonym for the underlined word.
   Pedestrians are not <u>allowed</u> to cross the road here.
   a) permitted
   b) watched
   c) known
   d) obliged

3. What's happened to Helmut? I haven't seen him . . . ages.
   a) during
   b) for
   c) while
   d) since

4. Find a synonym for the underlined word.
   This is a <u>safe</u> neighbourhood.
   a) local
   b) rich
   c) secure
   d) passive

5. A tiger is as dangerous . . . a lion.
   a) like
   b) than
   c) as
   d) then

6. Find a synonym for the underlined word.
   This version is able to carry a lot of <u>freight</u>.
   a) passengers
   b) plants
   c) fuel
   d) cargo

7. The basketball team . . . just around the corner.
   a) is
   b) has
   c) have
   d) is come

8. Fiona: 'I will drive to Hamburg on Friday.' Fiona said that she
   . . . to Hamburg on Friday.
   a) would drive
   b) drove
   c) drives
   d) will drive

9. Who permitted the pupils . . . school early?
   a) to leave
   b) who are leaving
   c) leaving
   d) leave

10. I would apply for the job, if I . . . you.
    a) were
    b) was
    c) had been
    d) have been

11. Jack hasn't got your money, Pamela hasn't got it . . .
    a) too
    b) either
    c) neither
    d) nor

12. She is thinking about . . . to Australia next month.
    a) will go
    b) go
    c) to go
    d) going

13. We don't have . . . business-class tickets.
    a) a few
    b) no
    c) much
    d) any

14. Find a synonym for the underlined word.
    What he said was quite <u>foreign</u> to our discussion.
    a) helpful
    b) connected
    c) alien
    d) instantaneous

15. Find a synonym for the underlined word.

    To have all your teeth out is a <u>certain</u> cure for toothache.

    a) stopping
    b) sure
    c) chosen
    d) recommended

16. Find a synonym for the underlined word.

    Max is the <u>eldest</u> of three brothers.

    a) first-born
    b) best educated
    c) fittest
    d) richest

17. Find a synonym for the underlined words.

    The old shop <u>merges into</u> the big foreign company.

    a) brings new ideas to
    b) suits
    c) is swallowed by
    d) reaches

18. Find a synonym for the underlined word.

    Outside the house it was frightening because of the <u>mist</u>.

    a) darkness
    b) jungle
    c) fog
    d) wind

19. Find a synonym for the underlined word.

    Reaching the <u>peak</u> of the mountain, they found a little hut.

    a) bottom
    b) other side
    c) top
    d) upper half

20. . . . . in history when remarkable progress was made within a
short span of time.
a) There have been periods
b) Periods have been
c) Throughout periods
d) Periods

MENTAL ARITHMETIC

As with many jobs, astronauts are often called
upon to do mental arithmetic. In most cases a
wrong answer will earn you nothing more than mild
embarrassment. However, there are some cases where
split-second calculations are required, and the lives of both you and
your crew will depend upon getting it right. For example, if your
main engine fails during the de-orbit burn on the return to Earth,
you have to make quick calculations regarding how long to burn
your secondary engines. Getting it wrong could mean re-entering
Earth's atmosphere at an angle that is too steep or too shallow, both
of which could have catastrophic consequences.

TEST 8

Try answering the following mathematical problems, without a
calculator or a pen and paper. You have ten seconds to answer each
question, and there is 1 point for each correct answer. Even if you
don't know the exact number, make your best guess.

Addition of numbers between 1 and 2000

$689 + 398 =$	$446 + 217 =$
$1115 + 21 =$	$251 + 897 =$
$1149 + 1992 =$	$1949 + 1040 =$
$1611 + 809 =$	$277 + 1849 =$
$1593 + 392 =$	$1912 + 646 =$

Subtraction of numbers between 1 and 2000

$1835 - 347 =$	$1977 - 839 =$
$1829 - 730 =$	$1096 - 163 =$
$1606 - 552 =$	$1845 - 462 =$
$1425 - 687 =$	$1849 - 830 =$
$1561 - 142 =$	$1575 - 360 =$

Multiplication of numbers between 1 and 20

$13 \times 8 =$	$3 \times 8 =$
$11 \times 15 =$	$6 \times 2 =$
$15 \times 18 =$	$3 \times 7 =$
$16 \times 18 =$	$12 \times 14 =$
$1 \times 9 =$	$18 \times 10 =$

Division of numbers between 1 and 100

Note: You should only answer with a rounded whole number (e.g. 5 for 4.8, or 4 for 4.4).

$34 \div 11 =$	$43 \div 8 =$
$85 \div 11 =$	$74 \div 8 =$
$93 \div 2 =$	$53 \div 5 =$
$44 \div 6 =$	$94 \div 8 =$
$39 \div 5 =$	$16 \div 5 =$

MEASUREMENT EXERCISES

Arithmetic skills on their own are not enough. An astronaut needs to be comfortable dealing with measurements of size, weight, speed and distance, along with percentages, ratios and fractions.

TEST 9

Try these exercises to see whether you're up to scratch. Can you find shortcuts to home in on which answer is correct, without needing to perform the full calculation? You get 1 point for each correct answer.

1. A student takes 15 seconds for a 100-metre sprint. How long does the student take if they run 2 km/h faster due to following wind?
   a) 13.8 seconds
   b) 13.6 seconds
   c) 13 seconds
   d) 14.2 seconds

2. A hotel charges an off-peak price of €98 for a single room with half-board. How much does the room cost during the main season, for which there is a 15 per cent extra charge?
   a) €110
   b) €112.70
   c) €114.40
   d) €85

3. A mother and her daughter are 60 years old, together. The mother is 36 years older than the daughter. How old are both persons?
   a) daughter 12, mother 48
   b) daughter 22, mother 38
   c) daughter 20, mother 40
   d) daughter 24, mother 60

4. A silver paperweight in the shape of a die has a side length of 5 cm. The mark '800' indicates that it is made of 800/1000 pure silver. Calculate what the silver is worth, if it has a density of 10.5 g / cm^3 and the silver price is €0.68 / g.
   a) €1229
   b) €654
   c) €714
   d) €701

5. A rectangle is 12 cm long and 8 cm wide. A second rectangle has the same area. How wide is the second rectangle, if it is 16 cm long?
   a) 4 cm
   b) 5 cm
   c) 5.5 cm
   d) 6 cm

6. A rectangular container with a side length of 3.5 metres and a width of 3 metres can take 21 m^3 of water. What is the height of the container?
   a) 3 metres
   b) 6 metres
   c) 2.5 metres
   d) 2 metres

7. A person wearing rollerblades takes 35 minutes to cover a distance of 12 km. What is their average speed in kilometres per hour?
   a) 20 km/h
   b) $20\frac{4}{7}$ km/h
   c) $21\frac{2}{7}$ km/h
   d) $20\frac{1}{7}$ km/h

8. A farmer from Texas has a square-shaped farm with an area of 100 km². How far must the farmer ride to check all of the farm's outer fence?
   a) 400 km
   b) 200 km
   c) 40 km
   d) 20 km

9. Two places (A and B) are 300 km apart. At 08:00 a freight train leaves A in the direction of B at 30 km/h. A passenger train leaves B in the direction of A at 11:00 at 60 km/h. Where and when do the two trains meet?
   a) at 13:00, 150 km from A
   b) at 13:40, 160 km from A
   c) at 13:20, 200 km from A
   d) at 13:20, 160 km from A

10. From a lottery prize Mr P gets one-fifth and Mrs N gets one-quarter. How much is the total lottery prize, if Mrs N gets €4000 more than Mr P?
   a) €80,000
   b) €54,000
   c) €36,000
   d) €140,000

11. At Frankfurt airport a passenger wants to get from Terminal 1
    to Terminal 2. On foot it would take them three minutes, on the
    conveyor belt two minutes. How long will it take them if they
    walk on the conveyor belt?
    a) 2 minutes
    b) 1.5 minutes
    c) 1.75 minutes
    d) 1.2 minutes

12. The reaction time of a driver is one second. How far does she
    drive on before hitting the brakes, if she drives at a speed of
    96 km/h and sees a red light?
    a) 26.7 metres
    b) 59.9 metres
    c) 29 metres
    d) 128 metres

13. The series $1 + \frac{1}{2} + \frac{1}{4} + \frac{1}{8} + \frac{1}{16}$ continues to $\frac{1}{512}$. What
    number is the series approximating to?
    a) 1
    b) 2
    c) 3
    d) 4

14. A staircase has eight steps, each 14 cm high. If the staircase is
    swapped for one with only seven steps, how high would each
    new step be?
    a) 16 cm
    b) 15 cm
    c) 15.6 cm
    d) 18 cm

15. Mining rubble can be removed by two lorries in eight days. To finish the work earlier an additional lorry is used, starting on the fourth day. How long does it take to clear the rubble?
    a) $7\frac{1}{3}$ days
    b) $6\frac{1}{3}$ days
    c) 6 days
    d) $6\frac{1}{4}$ days

16. Kerosene has a specific weight of 0.8. If an aeroplane has been refuelled with 4000 kg of kerosene, how many litres are inside its tank?
    a) 4000 litres
    b) 4800 litres
    c) 5000 litres
    d) 5600 litres

17. If $^4/_{12}$ of a cream cake are missing, the cake costs €24. How much does the whole cake cost?
    a) €40
    b) €36
    c) €20
    d) €48

18. Three sweets cost 15 cents. How many sweets could you buy with €1.65?
    a) 11 sweets
    b) 15 sweets
    c) 25 sweets
    d) 33 sweets

19. A keeper at a zoo has food that would last 78 days for seven lions. How long would the food last for 21 lions?
    a) 26 days
    b) 30 days
    c) 28 days
    d) 27 days

20. 145 grams of caviar costs €43.50. How much do 75 grams of caviar cost?
    a) €22.75
    b) €22.50
    c) €24.33
    d) €21.75

21. Oliver and Claire are ordering coffees totalling €72. For every three coffees Claire orders, Oliver orders five. The price for one coffee is €3. How much coffee did Oliver order?
    a) 7 coffees
    b) 10 coffees
    c) 15 coffees
    d) 12 coffees

22. A car with a turbo diesel engine uses 6 litres of diesel to travel 100 km. How much does it use for 250 km and how far can it go with 24 litres of fuel?
    a) 14 litres, 400 km
    b) 15 litres, 400 km
    c) 15 litres, 600 km
    d) 14 litres, 600 km

23. A wire of 48 cm length stretches to 52 cm if heated up. How long would a wire of 72 cm length become, if heated up?
    a) 78 cm
    b) 81 cm
    c) 77 cm
    d) 88 cm

24. If 4.5 metres of carpeting cost €90, how much would 2.5 metres cost?
   a) €50
   b) €42.50
   c) €45
   d) €40

25. How many points of intersection do five straight lines have at the most, in one plane?
   a) 2
   b) 5
   c) 10
   d) 4

26. The length of the sides of a die is 6 cm. The die weighs 0.54 kg. What is its density?
   a) 6 grams per cubic centimetre
   b) 3 grams per cubic centimetre
   c) 2.5 grams per cubic centimetre
   d) 2 grams per cubic centimetre

27. A line runs through the middle of a 3D shape such that it is equidistant from all surfaces. What is the shape?

      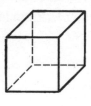   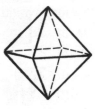

    a   b   c   d

   a) a sphere
   b) a cylinder (with no ends)
   c) a cube
   d) an octahedron

VISUAL MEMORY (CONTINUED)

Remember those shapes and numbers from page 25?

TEST 10

Without looking back at them, try and recall which numbers go with the following shapes. You get 1 point for each number you can remember.

1.

2.

3. **RC**

4.

5. **WT**

6.

7. **JF**

8.

9. **BZ**

10.

11. **KH**

12. **TH**

Answers

Test 1 (2 points)
a) + b) The dot ends up on the bottom both times.

Test 2 (20 points)

Test 3 (75 points)

Test 4 (12 points)

Test 5
1. d) Capacitor
2. b) 330 m/s
3. b) Copper
4. d) 220/230 volts
5. b) 4 volts
6. c) 2
7. a) A and B
8. b) Wheel 2
9. a) A
10. a) A
11. c) 1 and 3
12. a) 2. This gives the wheel the most room to swivel around the rotation axis.
13. a) 3
14. c) intake – compression – ignition – exhaust
15. a) Reduction of the engine temperature
16. c) 4–6 litres
17. a) The maximum permitted rpm
18. d) Kerosene
19. b) 1.8–2.8 bar
20. a) Calvin
21. a) Increases with voltage
22. c) It is reduced

SURVIVAL TRAINING

On 19 March 1965, the Russian Voskhod 2 spacecraft was attempting to re-enter Earth's atmosphere when the primary retrorockets failed. The crew, including Commander Pavel Belyayev and Alexei Leonov, fortunately landed safely, but over 300 km off course in a forest full of wolves and bears. As night fell the temperature dropped to −30°C. The next day a rescue team had to ski to the landing site with food, water and wood to make a fire, while others chopped down trees to clear an area for a helicopter to land. The crew were eventually rescued, but events could have been much worse.

The story of Voskhod 2 is a reminder that space travel doesn't always go to plan. Re-entry in particular can go wrong, and descent modules can land in dangerous environments. That's why once you have been selected to be an astronaut you will be trained in survival skills. You have to be prepared for a landing anywhere on the surface of the Earth, including sea, desert, rainforest or glacier.

In 2010, my five fellow astronaut trainees and I – the astronaut class of 2009 – spent two weeks living in the wilderness. We were thrown out of a helicopter, left to drift at sea and abandoned under the hot Mediterranean sun. Although there were some instructors around, we largely had to fend for ourselves.

During survival training, astronauts develop other vital skills, including climbing and descending cliffs, crossing rivers and navigating by the stars. My class also learned to administer medicine and first aid, and were taught to use and repurpose the hardware aboard a spacecraft for survival. During one exercise, as we didn't have a spacecraft to hand, we stripped down an old car, using its parts to build a shelter, to make traps and snares to catch food and fend off hostile animals, and to fashion equipment for fishing.

Once you are assigned to a mission as an astronaut, you start learning more detailed skills related to specific space vehicles and environments. If you are flying on the Soyuz spacecraft, as I did to get to and from the ISS, you will visit the Yuri Gagarin Cosmonaut Training Center near Moscow for winter survival training.

A mock-up of the Soyuz descent module was positioned in the snow and we had to practise changing into thermal suits to keep us warm. We were wearing five layers of clothing in total. That was very much needed because it was −24°C outside. We then felled some trees to build a fire and construct a makeshift shelter using the parachute from the Soyuz. We had to survive out there for two nights – the same amount of time Leonov and Belyayev were stranded in 1965.

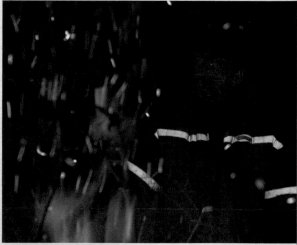

Survival training also sees astronauts practise for a Soyuz landing at sea. This is unusual because the spacecraft is supposed to come down on land, but we have to know what to do if it ends up being a splashdown instead. Over seventy per cent of Earth's surface is covered in water after all, and in the event of a rapid emergency descent from space we may not get much choice of landing site.

One of the greatest challenges was simply changing clothes in a very cramped environment. The crew have to change from pressurised spacesuits to full thermal protective clothing and immersion suits to help us survive in the water. All of this is completed inside the tiny Soyuz descent module, with the hatches closed and the spacecraft bobbing on the water. To add to our discomfort, this exercise is conducted in the height of summer. The temperature in the capsule rapidly climbed to around 32°C and we were in there for nearly an hour. In that time I lost nearly 2.5 kg of my body weight just through sweating. Needless to say it was a huge relief to finally exit the capsule and hit the cool water.

CAVES

As you enter the Sa Grutta caves in Sardinia, the sunlight starts to seep away. Soon the only illumination will come from the torch on your hard hat. This maze of tight tunnels and cathedral-sized chambers is your home for the next six days. You are here as part of your mission training, to take part in the Cooperative Adventure for Valuing and Exercising human behaviour and performance Skills (CAVES) programme.

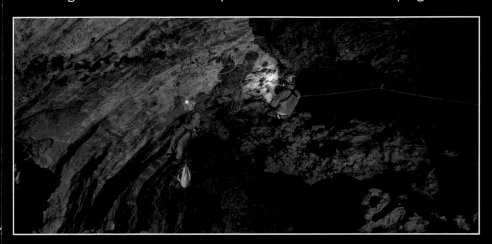

It might not be immediately obvious why a Mediterranean cave system is a good place to train for a trip to space. However, there are many similarities to life in orbit. In the cave you have to work as a team to navigate a series of vertical climbs, descents and traverses using steel cables and climbing ropes. This is like navigating your way along the outside of the International Space Station on a spacewalk. In both situations you have to remember your way back to safety, whether that's back to the airlock in space or to the mouth of the cave. You are a long way from help if anyone slips and has an injury. You have to maintain constant vigilance to remain attached to safety mechanisms. A caving suit is necessary to protect you, just like a spacesuit.

The isolation experienced during CAVES training is eerily similar to life in space, as communication with the outside world is limited to two radio broadcasts a day. It is easy to lose your sense of time while cut off underground, and you have to rely heavily on your watch. This is a good match to conditions on the ISS where its rapid orbit around the Earth means you experience a sunrise or sunset every forty-five minutes, making an artificial sense of time crucial to staying on schedule without the natural cues you are used to.

What you are really in the caves for is to learn about teamwork, conflict resolution and decision-making in a challenging environment where risks are commonplace and safety is paramount. It is essentially a masterclass intended to develop the skills covered in an astronaut's Human Behaviour and Performance (HBP) training (see page 98).

During the CAVES training, your fellow 'cavenauts' often come from different space agencies around the world, meaning there is a variety of cultures and approaches. This is partly to replicate working conditions on board the ISS with its international crew. Loredana Bessone, an instructor who oversaw my CAVES training, says, 'Imagine you are going on a six-month cruise with a group of people you have just met at work. After a week you start to get annoyed by their tutting or their habit of leaving their socks around.' CAVES offers a real-world chance to learn how to iron out these difficulties in a place where crew safety is on the line and privacy is non-existent. Bessone says, 'You have to develop a common culture, to learn from mistakes and grow as a team.'

Cosmonaut Sergei Korsakov, ESA astronaut Pedro Duque, taikonaut Ye Guangfu, Japanese astronaut Aki Hoshide and NASA astronauts Ricky Arnold and Jessica Meir explore the caves of Sardinia.

One way to achieve team cohesion during CAVES training is to work on both verbal and non-verbal communication. In the same way that a car has an indicator to show other drivers what action it is about to take, providing similar cues to your fellow cavenauts, or astronauts in space, will ensure operations run more smoothly. During caving, as you get to know your crewmates, you are encouraged to look for behaviours that depart from the norm. If a loud person suddenly goes quiet, or a softly spoken colleague starts raising their voice, it can be an early indicator of an issue. Better to get ahead of the situation before it escalates into something that could put the crew's safety in doubt. Looking for these little clues is all part of an astronaut's situational awareness training (see page 135). It is also an invaluable part of conflict management; spot a possible clash point early and you can tackle it before it balloons into a bigger problem. This team cohesion is critical to the daily operation of the ISS.

TEAMWORK

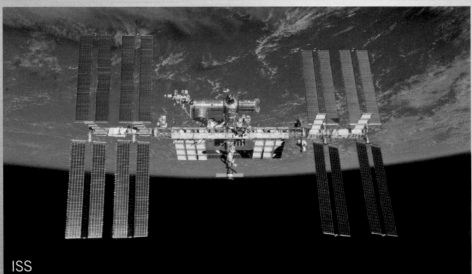

ISS

23. d) can be measured in Fahrenheit
24. a) Direct and alternating
25. b) Ampere
26. d) Resistance multiplied by current
27 c) The positive and negative poles of a voltage source have a direct connection (= 0 ohm)
28. a) A shows the best gate design. Bracing should always 'lean into the opening', as the weight of the gate will act downwards at the side furthest from the hinge.

Test 6 (16 points)

1. Side, Above, Below, Above, Side, Side, Side, Above
2. Side, Side, Side, Above, Above, Side, Below, Below

Test 7

1. a) hasn't
2. a) permitted
3. b) for
4. c) secure
5. c) as
6. d) cargo
7. a) is
8. a) would drive
9. a) to leave
10. a) were
11. b) either
12. d) going
13. d) any
14. c) alien
15. b) sure
16. a) first-born
17. c) is swallowed by
18. c) fog
19. c) top
20. a) There have been periods

Test 8

Addition of numbers between 1 and 2000

$689 + 398 = 1087$

$1115 + 21 = 1136$

$1149 + 1992 = 3141$

$1611 + 809 = 2420$

$1593 + 392 = 1985$

$446 + 217 = 663$

$251 + 897 = 1148$

$1949 + 1040 = 2989$

$277 + 1849 = 2126$

$1912 + 646 = 2558$

Subtraction of numbers between 1 and 2000

$1835 - 347 = 1488$

$1829 - 730 = 1099$

$1606 - 552 = 1054$

$1425 - 687 = 738$

$1561 - 142 = 1419$

$1977 - 839 = 1138$

$1096 - 163 = 933$

$1845 - 462 = 1383$

$1849 - 830 = 1019$

$1575 - 360 = 1215$

Multiplication of numbers between 1 and 20

$13 \times 8 = 104$

$11 \times 15 = 165$

$15 \times 18 = 270$

$16 \times 18 = 288$

$1 \times 9 = 9$

$3 \times 8 = 24$

$6 \times 2 = 12$

$3 \times 7 = 21$

$12 \times 14 = 168$

$18 \times 10 = 180$

Division of numbers between 1 and 100

Remember all answers need to be rounded to the nearest whole number.

$34 \div 11 = 3$

$85 \div 11 = 8$

$93 \div 2 = 47$

$44 \div 6 = 7$

$39 \div 5 = 8$

$43 \div 8 = 5$

$74 \div 8 = 9$

$53 \div 5 = 11$

$94 \div 8 = 12$
$16 \div 5 = 3$

Test 9
Measurement Exercises
1. a)   13.8 seconds
2. b)   €112.7
3. a)   daughter 12, mother 48
4. c)   €714
5. d)   6 cm
6. d)   2 metres
7. b)   20⁴⁄₇ km/h
8. c)   40 km
9. d)   at 13:20, 160 km from A
10. a)   €80,000
11. d)   1.2 minutes
12. a)   26.7 metres
13. b)   2
14. a)   16 cm
15. b)   6⅓ days
16. c)   5000 litres
17. b)   €36
18. d)   33 sweets
19. a)   26 days
20. b)   €22.50
21. c)   15 coffees
22. b)   15 litres, 400 km
23. a)   78 cm
24. a)   €50
25. c)   10
26. c)   2.5 grams per cubic centimetre
27. b)   a cylinder (with no ends)

HARD SKILLS DEBRIEF

How did you get on with the selection questions? Add up all your points and compare your performance with the scores given below.

0–100

Not everyone is cut out to be an astronaut and perhaps your strengths lie elsewhere. There's no doubt you can improve your score and certain skills and techniques can be developed over time, but lot of the ability to answer questions like this is hard-wired into our brains. What did you struggle with most?

101–150

There's work to be done, but don't despair. Persevere, work on those areas you found difficult, and you may be able to improve your score.

151–200

Excellent job. You have many of the qualities that selectors look for when searching for good astronaut candidates. If you fell down in one particular part of the test, then practising that skill or brushing up on those facts will really help.

201–240

Incredible! You've reached the dizzy heights of the top of the pack. To get a score like this, you must be a calm all-rounder with some serious skills. Well done.

SPEED VS ACCURACY

When you tackled the maths questions, I gave you a time limit of ten seconds to answer each one. We also had a time limit in the selection process, and it led to an interesting choice between whether you thought the selectors were after speed or accuracy. Did you decide to choose one over the other? In truth, the tests are looking for both – speed is no good if you often get the answer wrong; a slow but correct approach isn't ideal, either, if you cannot complete a large part of the test. A balanced technique is the optimum approach.

PART TWO
ASTRONAUT REQUIREMENTS

THE APPLICATION

Well done on completing the first round of hard skills. Now we turn to the application form itself. This chapter outlines the wide range of personal attributes that you will need to possess in order to be selected as an astronaut, including medical, psychological and professional qualifications. The criteria are strenuous, but you may also be surprised by some of the requirements.

OPEN-ENDED QUESTIONS

Let's start by answering the following questions. Your answer to each question should be no more than 750 words in length.

1. Why do you want to become an astronaut?
2. In your opinion, what are the main tasks that should be performed by an astronaut?
3. Write a candid description of yourself as a person (including your assets and your limitations).

When I applied to join the ranks of the European Space Agency in 2008, the three questions above were the final questions that I had to complete on my ESA Astronaut Selection Test application form. The full form is 15 pages long, and much of it is dedicated to a candidate's professional experience and qualifications, to ensure you meet the exacting criteria, which we will discuss in a moment. But open-ended questions, such as the ones above, are designed to differentiate between candidates with similar professional backgrounds, and to get an early sense of your personality and psychological profile.

Before you answer the questions above, think very hard about them and spend some time carefully examining your responses. I certainly did. I remember I took hours over it. What life experiences would you draw on? What character traits would you emphasise, which your friends, family members and colleagues admire in you? Ultimately, how can you distil your excitement for the job, your experience and your personality into just a few paragraphs?

Below are three more questions from the application form. How you would answer these questions, using no more than 25 words for each answer? At the end of this section we will revisit all of these questions, to hear from an astronaut instructor about what ESA is really looking for in a candidate's answers.

4. If appointed, would you agree to take up residence in the Cologne region (Germany) – headquarters of the European Astronaut Centre (EAC)?
5. Would you be willing to live in Star City, Russia, for an extended period (two to three years) – location of the training site for astronauts?
6. How would you describe your manual skills (regarding repairing, disassembling and reassembling equipment, etc.)?

These questions may seem ordinary on the surface, but navigating the application process is one of the hardest obstacles to overcome on your way to becoming an astronaut – only a small percentage make it through even this early stage. When I applied in 2008, my application was one of 8172 suitable applications received by ESA. Gerhard Thiele is a former ESA astronaut and was the project manager responsible for that selection process. He explained that the applications were considered by a large team of ESA Astronaut

Selection Test recruiters, who sat together in the same room to allow easy discussion of the merits of each candidate. Applications were marked 'yes' or 'no' and were then placed in different piles. All those marked 'no' were reviewed again by a separate team, and two 'no's would see you eliminated from the process.

Of the 8172 candidates who applied, only 918 were selected to move on to the next round of testing for 'hard skills'. That is a pass rate of just 11 per cent; so 89 per cent of candidates failed to get through. It is an important lesson on your journey to becoming an astronaut that sometimes you don't get through and you have to try again. NASA astronaut Clayton Anderson was famously rejected a staggering 14 times, before his 15th application was finally successful. This section will outline the application process, to improve your chances.

ASTRONAUT REQUIREMENTS

Apart from open-ended questions, the majority of the application form is devoted to assessing a candidate's suitability in meeting the stringent requirements for space. The astronaut selection process is based on recruiting individuals who can meet the needs of various space missions, both current and foreseen. Major aspects taken into consideration are psychological suitability, scientific and technical competence and fulfilment of medical criteria. Not everyone will meet these criteria; and some, such as height and eyesight, are beyond your control. However, in order to give yourself the best chance of your application form making the 'yes' pile, you will need to meet ESA's specific requirements. Here is an example of what was required during the 2008 selection process.

GENERAL REQUIREMENTS

Applicants, male or female, must be nationals of an ESA Member State: Austria, Belgium, the Czech Republic, Denmark, Estonia, Finland, France, Germany, Greece, Hungary, Ireland, Italy, Luxembourg, the Netherlands, Norway, Poland, Portugal, Romania, Spain, Sweden, Switzerland and the United Kingdom.

The preferred age range is 27–37, and applicants must be within the height range of 153–190 cm. You must speak and read English and have a university degree (or equivalent) in Natural Sciences, Engineering or Medicine, and preferably have at least three years' postgraduate professional experience in a related field. Flying experience is welcome.

MEDICAL REQUIREMENTS

Applicants should be in good health, have a satisfactory medical history, be of normal weight and have a sound mental disposition. Specific tests later in the selection process will be performed to evaluate applicants' bodily systems (muscular, cardiovascular and vestibular). Once selected, during training astronauts will be expected to use facilities such as centrifuges, rotating chairs, pressure chambers and aircraft, which will push an astronaut's body to its limits.

PSYCHOLOGICAL REQUIREMENTS

The general characteristics expected of applicants include good reasoning capability and memory, concentration, aptitude for spatial orientation and manual dexterity. An applicant's personality should be characterised by high motivation, flexibility, gregariousness, empathy with fellow workers, emotional stability and a low level of aggressiveness. For long-term flights on the International Space Station, the ability to work as a team member in an intercultural environment is of high importance.

PROFESSIONAL REQUIREMENTS

The candidate should be knowledgeable in scientific disciplines and should have proven outstanding ability in applicable fields, preferably including operational skills. This doesn't mean that you have to be a scientist; in fact astronauts are selected from many different backgrounds. However, science will play an important role in both training and spaceflight, so a basic knowledge of and keen interest in science are essential.

FREQUENTLY ASKED QUESTIONS

As an astronaut candidate, if you have any follow-up queries related to the requirements given above (and you are likely to), ESA provides the following helpful information, to assist you with the application form.

Q *Which medical and psychological standards will be used to select the candidates?*

A In general, normal medical and psychological health standards will be used. These standards are derived from evidence-based medicine, verified by clinical studies. *Note*: If you are successful in your application, you will undergo medical tests later on in the astronaut selection process.

- An applicant should be able to pass a JAR-FCL 3, Class 2 medical examination or equivalent, conducted by an Aviation Medical Examiner certified by his/her national Aviation Medical Authority. (*Note*: This is a medical examination typically used to test pilots.)
- The applicant must be free from any disease.
- The applicant must be free from any dependency on drugs, alcohol or tobacco.
- The applicant must have the normal range of motion and functionality in all joints.
- The applicant must have visual acuity in both eyes of 100 per cent (20/20), either uncorrected or corrected with lenses or contact lenses.
- The applicant must be free from any psychiatric disorders.
- The applicant must demonstrate cognitive, mental and personality capabilities to enable him/her to work efficiently in an intellectually and socially highly demanding environment.

Q *Do I need to be fit to become an astronaut? Which sport should I pursue?*

A It is important to be healthy, with an age-adequate fitness level. Astronaut selectors are not looking for extreme fitness or top-level athletes – too many over-developed muscles may be a disadvantage for astronauts in weightlessness. There is no specific sport that can be recommended.

Q *Do astronauts develop serious health problems during their stays in space?*

A No, there are no dangerous conditions that develop because of spaceflight. However, the space environment is hazardous, and the astronauts' well-being depends on life-support systems. Weightlessness does have potentially temporary negative effects on human physiology, such as physical deconditioning and bone demineralisation. The ESA Crew Medical Support Office and its staff are responsible for avoiding such hazards and preventing the space environment from affecting the physical and mental health of the astronauts. The environment and life-support systems are closely monitored, and there is a thorough preventive and countermeasure programme.

Q *Is it more difficult for a woman to become an astronaut than a man?*

A No. The medical and psychological requirements for women and men are identical, apart from some gender-specific medical examinations.

Physical fitness and cardiovascular fitness are always evaluated on an individual basis, and the fitness target values are adjusted to the physiological differences between men and women.

Q *My vision is not perfect; can I still become an astronaut?*

A There is no clear yes/no answer, because there is such a multitude of visual defects. However, vision problems account for most medical disqualifications. The main tests involve visual acuity, colour perception and 3D vision.

Wearing spectacles (glasses) or contact lenses is not a reason for disqualification per se, but it has to be evaluated if, for example, a visual defect is known to progress rapidly. This could mean disqualification. Minor visual defects, even though requiring lenses, may be regarded as compatible with space duties.

Recently a variety of surgical interventions to correct visual acuity has become more common. Some of these procedures will lead to disqualification, while others are acceptable. Every case will be judged individually.

TEST DEBRIEF

If you fulfil most of the above criteria, congratulations! Your application stands a great chance of progressing to the next stage. Before we move on, let's return to the six open-ended questions posed on the application form:

1. Why do you want to become an astronaut?
2. In your opinion, what are the main tasks that should be performed by an astronaut?
3. Write a candid description of yourself as a person (including your assets and your limitations).
4. If appointed, would you agree to take up residence in the Cologne region (Germany) – headquarters of the European Astronaut Centre (EAC)?
5. Would you be willing to live in Star City, Russia, for an extended period (two to three years) – location of the training site for astronauts?
6. How would you describe your manual skills (regarding repairing, disassembling and reassembling equipment, etc.)?

Gerhard Thiele's view on what makes a good answer to Question 1 is instructive and somewhat surprising. 'We are not looking for a specific type of response,' he says. 'There is no right or wrong answer.' ESA does, however, use the candidate's answer as an opportunity to assess a range of criteria.

First and foremost is attention to detail, which is a crucial quality to possess if you are an astronaut. 'We have had people who write five pages about why they want to be an astronaut,' Thiele says. 'Despite their obvious enthusiasm, these candidates are not considered, because the question specifically asks for a short answer (750 words maximum).' In space, getting the small details correct is integral to the success of each mission, and written instructions or messages provide a key role. 'Sweat the small stuff' is a well-known astronaut mantra. Identifying and studying the small details related to a task will give you options when big, life-threatening events occur.

The reason ESA asks for a short answer is connected to the next criterion for which they are testing – communication skill. As we examine during astronaut training in Part Three of the book, clear and concise communication is a key skill for astronauts. Whether communicating over the radio to Mission Control during a spacewalk or conducting a TV interview back on Earth, astronauts must be able to share complex information precisely and with brevity. If you are verbose in a critical situation, your failure to communicate vital information in a timely manner could have drastic consequences. Equally, if you cannot share the wonders of space readily and in an inspiring fashion with the general public during interviews, you will have missed an opportunity to fire the imagination of the next generation of space explorers.

English skills of spelling and grammar are also tested in these questions. This is particularly relevant to astronauts who are applying with English as a second language, but it is just as important for native speakers. Answers that include typos and multiple errors suggest that the candidate has not checked their answers thoroughly, or is careless with detail.

For Question 2, ESA does not require the candidate to address all the possible roles of an astronaut, as the job has so many different responsibilities. It does, however, expect the candidate to demonstrate that they have researched the role widely and are able to share their thoughts in an impactful, succinct manner.

For Question 3, when you are asked to write a candid description of yourself, ESA anticipates an honest answer that shares both your strengths and your weaknesses. Can you demonstrate that you are qualified and confident in your abilities, yet humble and without ego at the same time? A balanced response might suggest that you have the psychological make-up to be both a leader and a team player, as both skills are required for working in space.

Questions 4 and 5, on the face of it, have straightforward binary answers. As part of the training process, astronauts are required to move to Cologne to train at the EAC and subsequently to Star City in Russia. If a candidate says 'no', they are likely to be rejected. But the astronaut selectors are interested, from a psychological point of view, to see if candidates will simply accept these lifestyle choices, or if they will consider the disruptive impact that they will potentially have on their lives and their family's lives. The latter type of response, which may agree to relocation to those cities, but ask how ESA will support their family during this process, suggests that a candidate is responsible and has the confidence to question authority. These are skills that may need to be put to the test in space, when Mission Control occasionally gives an astronaut the wrong instruction or piece of information.

Astronauts must have 'manual' skills and must possess a practical DIY mindset in order to repair, clean or assemble parts of the space station. Question 6 is asking not just for a list of the candidate's skills, but for their own assessment of their skills. Astronaut instructors at ESA will be intrigued to read whether the candidate includes their shortcomings as well as their strengths. During astronaut selection there is often a psychological component, even when the question appears simple.

We have focused in this section of the book on psychological and medical questions, because they are a crucial part of the astronaut selection process. Due to the mental and physical exertions of a mission to space, the mind and body need to be in excellent shape. Unfortunately, the tests are so stringent that nearly half of all candidates will not progress for medical reasons. If you apply and this happens to you, it can be very disappointing, but don't lose heart. The career of one of my colleagues, Hervé Stevenin, is a great example of this. I would like to finish this chapter by briefly examining his own space journey.

Hervé has space in his bones. Ever since he was a child he wanted to become an astronaut, but he missed out a number of times. Born in France in 1962, he was too young when the French selected their first batch of space travellers in 1985. At the time, he was studying engineering, but the selectors were looking only for graduates with at least a year's work experience. So instead he set about learning as much as he could to prepare himself for the next selection round, spending time in aircraft, learning to dive and gaining valuable parachuting skills. He soon had all the qualities they had been looking for in 1985. Yet when the next French selection rolled around in 1990, the goalposts had changed and they were looking for test pilots to make the transition to astronaut. He missed out again.

There wouldn't be another selection process until the 2009 European Space Agency intake, when he applied alongside me and my classmates. Hervé had a massive advantage over most of us: he had joined ESA as part of the astronaut training team. Before that he had trained French astronauts for their trips to the (now defunct) Russian Mir space station. He even received spacewalk training at NASA's Johnson Space Center in the Neutral Buoyancy Laboratory in Houston in 2004.

Hervé made it through the gruelling rounds of psychological testing, and was one of the remaining forty-five candidates sent for medical tests. If he could get through that stage, he would be on to the final interviews. But then the hammer blow struck. 'They found a very small medical condition during a test,' he says. That wouldn't have been a problem if the selection was for short-duration missions as it had been in 1985 or 1990, but the missions would be six-month stays on the International Space Station. That long in microgravity plays havoc with your body, causing changes to the cardiovascular system, the immune system and to muscles and bones that atrophy. Hervé's existing condition made him too much of a medical risk and he was out.

'When you go through a selection process and you get kicked out by something beyond your control, when you know that it's your last chance because next time you'll be too old, that's difficult. You have a bad time, I can tell you.' It was particularly tough because Hervé then had to train the candidates who were eventually selected. 'I made peace with it,' he says. 'You find the resources inside you to do it.'

Hervé has made the most of the unique opportunity to work with astronauts from all over the world. He is the only person in Europe, aside from astronauts themselves, who has trained in both the NASA and Russian EVA spacesuit, training with the latter at Star City outside Moscow. He's been part of a NASA Extreme Environment Mission Operations (NEEMO) mission to re-create the environment of space at the bottom of the ocean (see Photo Inset 2). You'll also find him training astronauts on the zero-g 'Vomit Comets' (Photo Inset 2), and three times he's been part of the search and rescue teams in Kazakhstan helping to recover astronauts on their return from space in the Russian Soyuz capsule.

Sometimes you can work your entire life towards a single goal and for one reason or another fail to make it. What you do with that disappointment is crucial. Hervé has turned it into something positive and he is an invaluable part of the European Astronaut Centre team working to push the boundaries of human spaceflight. For almost every ESA astronaut who spacewalks and looks back on Earth, Hervé has been the one who trained them. So never give up.

FINAL SELECTION

The psychological questions we have explored in this chapter inform the final stage of the selection process – interviews with senior members of the European Space Agency staff. Candidates who have passed the medical tests are now quizzed on their interpersonal skills, their temperament for space and their life ambitions, as well as factors such as their ability to speak to the media – astronauts have to be excellent champions of science and human spaceflight, both on Earth and in space. This is a nerve-wracking process, often with one candidate sitting in front of a desk of seasoned astronauts, astronaut trainers and senior management.

It's not easy, and I remember answering several challenging questions, ranging from the purpose of human spaceflight and my knowledge of scientific accomplishments, to analysing my personal strengths and weaknesses.

But in this final stage it's important to remember that you already have what it takes to be an astronaut. What the panel are looking for is what sets you apart from the other candidates. If you can be yourself, and allow your personality and character to shine through, there's a good chance you will prevail.

In the next section of the book, you will join the latest astronaut class and begin your basic training, where you will be tested on a wide range of personal 'soft skills' among many other attributes. You have joined the elite – in the 2008 ESA selection, only six astronauts were chosen from the 8172 candidates who applied. Your path to space is getting closer than ever. Some of the tests and challenges are about to get even harder, but I loved every minute of it!

PART THREE
ASTRONAUT TRAINING

BASIC TRAINING

Well done! You have risen above the competition and have officially been selected to join the latest astronaut class of the European Space Agency.

I heard this news via a phone call on a warm spring evening in 2009. In an instant my life was transformed and a flood of emotions washed over me. Mixed with pure excitement and elation were feelings of apprehension, and an abundance of questions to which I had no answers. I was about to move my family abroad to Germany and embark on one of the toughest training courses ever devised. I knew very little about what life as an astronaut would be like: would I make the grade? Even if I did, there was no guarantee of securing a mission to space. But if life has taught me one thing, it is that you have to grasp opportunities when they arise – and this was a once-in-a-lifetime opportunity that was not going to pass me by.

On your journey to becoming an astronaut the hard work really begins. You are about to start 18 months of intensive basic training, to get you up to speed with all the skills you will need to survive and thrive in space. Then it is on to advanced training (see Photo Insets 1 and 2), in readiness for the launch.

How will you fare, learning new languages, working on your interpersonal skills, developing your robotics techniques and perfecting your concentration to perform a spacewalk? Before we look at exactly what is involved in both basic and advanced training, try the following tests, which are designed to establish your aptitude for languages and your ability to handle difficult interpersonal situations.

LANGUAGE APTITUDE TEST

One of the hardest parts of basic training, for me at least, was learning Russian – the other official language (besides English) spoken on the International Space Station. In fact three of the first six months of your astronaut training will consist of intensive Russian lessons, including a month-long immersion trip, spent living with a Russian family. It's crucial to have a good grasp of the language, not least because everything written inside the Soyuz spacecraft that you may fly to the ISS is in Russian. This includes the instruments, the control panels and the flight documentation (there is no English translation). You will speak to Moscow's Mission Control Centre exclusively in Russian and, once you arrive on the ISS, you will speak to your fellow Russian cosmonauts on a daily basis. In addition to these operational reasons, there is a social dimension to knowing the language. As an ESA astronaut, you will spend many months training in Russia with your fellow cosmonauts and, in order to bond with them properly, understanding their language is key.

I found learning Russian particularly difficult, because I am not a natural linguist. When I was training on the Soyuz simulator in Star City, a Russian reporter asked me a question in Russian at a press conference. I had to answer in Russian, in front of the assembled media – a nerve-racking moment, and I'm still not sure she understood what I said! I'd learnt how to speak technical Russian, how to read procedures and checklists in Russian and how to follow Russian commands given over the radio, but I still struggled with conversational Russian. I worked hard to brush up on it ahead of launch day.

Learning Russian is difficult because you have to contend

with a different alphabet – Russian is written in Cyrillic. At first it may seem completely alien, but Cyrillic has some similar letters to the English alphabet, just with different sounds associated with them. When learning a different language, it helps to have a certain mindset. You need to have a flexible approach, look for patterns and concepts that are similar to your native language, and try to associate new words with something familiar, to help you memorise vocabulary.

Below is a series of tests designed to see how well you can adapt to learning a new language. Good luck! Or, as we say in Russian, Удачи!

TEST 1

Without using a dictionary, translate the following Russian space-related words into English. Use the following chart to help you.

А а	a as in father	**К к**	k as in class	**Х х**	h as in loch
Б б	b as in but	**Л л**	l as in love	**Ц ц**	ts as in its
В в	v as in van	**М м**	m as in mother	**Ч ч**	ch as in chess
Г г	g as in get	**Н н**	n as in name	**Ш ш**	sh as in fish
Д д	d as in dress	**О о**	o as in bottle	**Щ щ**	shch as in fresh chat
Е е	ye as in yesterday	**П п**	p as in paper	**Ъ ъ**	'hard sign'
Ё ё	yo as in yonder	**Р р**	r as in error	**Ы ы**	i as in bill
Ж ж	zh as in measure	**С с**	s as in smile	**Ь ь**	'soft sign'
З з	z as in zoo	**Т т**	t as in ten	**Э э**	e as in bet
И и	ee as in meet	**У у**	u as in cool	**Ю ю**	yu as in Yugoslavia
Й й	y as in toy	**Ф ф**	f as in farm	**Я я**	ya as in yard

- модуль
- станция
- космонавт
- ракета
- капсула
- парашют
- процедура
- орбита
- камера
- траектория

TEST 2

Below are eight words in Russian that you would normally associate with each other. They have been rearranged. Using the chart given on page 85, first translate the words into English. Then put them into the correct order.

- Сатурн
- Меркурий
- Юпитер
- Нептун
- Марс
- Венера
- Земля
- Уран

Now you must examine some other languages. The following exercises will still test your core language-learning abilities and will also show you some of the other languages spoken by ESA astronauts and the international community of space.

TEST 3

Match the following eight Dutch[1] words for animals with their English equivalents. For a bonus point, name the animals that have been to space, from the following list. Before humans, animals were some of the first space explorers on early missions.

Dutch
- vlieg
- hert
- haan
- schildpad
- kikker
- aap
- neushoorn
- spin

English
- monkey
- deer
- frog
- rhinoceros
- turtle
- spider
- cockerel
- fly

---

1 ESA's largest European office, the European Space Research and Technology Centre (ESTEC), is situated in the Netherlands.

TEST 4

Below is a list of German words related to food. By thinking of similar words in English, and/or in other languages that you may already know, your task is to give a definition of each word – your answers should be a single word or short phrase. For added difficulty, there are five words hidden in the list that are not foods. Can you identify them?

- Reis
- Suppe
- Zucker
- Brezel
- Tischdecke
- Wassermelone
- Mais
- Salatgurke
- Honige
- Mittagessen
- Nudeln
- Schweinekotelett
- Karotten
- Gabel
- Becher
- Krabbe
- Auster
- Apfel
- Speisekarte
- Zitrone

TEST 5

The list below is composed of common space-related words or phrases. They are written in the native languages of the ESA astronauts who were selected alongside me in the class of 2009. Your task: to match each word or phrase with the English equivalent.

Hint: Thinking about similar words in English may help. For a bonus point, can you identify the different languages in the first list?

- passeggiata nello spazio
- amarrage
- hjelm
- fehlfunktion
- vaisseau spatial
- brûler le moteur
- rifiuto spaziale
- luftsluse
- densità ossea
- sonnenkollektor

English
- malfunction
- bone density
- spacecraft
- airlock
- space walk
- docking
- helmet
- solar panel
- engine burn
- space junk

TEST 6

Here is a sentence in Italian: 'Ho pulito il bagno dello spazio.'
It means: 'I cleaned the space toilet.'

Using the word and sentence structure above as a guide, put the following words into the correct order, to form another space-related sentence in Italian: 'Pannello riparato ho solare il.'

For a bonus point, translate the new sentence into English.

TEST 7

The following sentences are facts about space written in French. Your task is to deduce the meaning of each sentence.

Hint: Again, think about similar words in English.

1. Douze astronautes ont marché sur la lune.
2. Dans l'espace tu flottes.
3. Vous voyagez dans l'espace à bord d'une fusée.
4. Les astronautes réparent la station spatiale pendant les sorties dans l'espace.
5. La station spatiale internationale est une collaboration entre de nombreux pays.

TEST 8

Can you decipher Chinese?

Some ESA astronauts now learn Chinese as well as Russian. ESA and Chinese astronauts are beginning to train together, with the ultimate goal of working together on the future large modular Chinese space station.

If Russian is a hard language to learn, then Chinese is arguably even tougher, because it uses a writing system different from English and different even from the Cyrillic alphabet. However, the way in which the Chinese characters are constructed sometimes gives you a visual clue as to what a word means.

Look carefully at the **bold** elements of the five Chinese words given below. Your task is to match them to these equivalent English words:

- window
- plane
- rocket
- astronaut
- chair

火箭

椅

平面

宇航**员**

窗**口**

TEST 9

Reflecting the international community of space, your next task is to find the appropriate astronaut/space-related words that have the same meaning in different languages. For each of the five English words given below, find two foreign words with the same meaning from the list underneath.

- **Group 1:** buoyancy
- **Group 2:** teamwork
- **Group 3:** training
- **Group 4:** parachute
- **Group 5:** constellation

Foreign words: sternbild, flottabilité, entraînement, fallschirm, stjörnumerki, opdrift, lavoro di squadra, samenspel, formazione, valskerm

TEST 10

Having good language skills doesn't simply mean learning foreign words. Space travel is full of technical language with precise scientific meanings. Examine the words below and provide a short definition – in as few words as possible. Even if you don't think you know the exact definition, good linguists will be able to use similar words they already know to make a best guess. How close can you get?

- ablation
- antipodal
- ephemeris
- geostationary
- gimbal
- hypergolic
- meridian
- perigee
- radiometer
- transponder

Answers

Check your answers below and give yourself a score, based on the points indicated.

Test 1
Russian words (10 points):
module, station, cosmonaut, rocket, capsule, parachute, procedure, orbit, camera, trajectory

Test 2
Russian sequence (8 points):
Mercury, Venus, Earth, Mars, Jupiter, Saturn, Uranus, Neptune

Test 3
Dutch animals (8 points):
- vlieg = fly
- hert = deer
- haan = cockerel
- schildpad ('shield toad') = turtle
- kikker = frog
- aap = monkey
- neushoorn ('nose horn') = rhinoceros
- spin = spider

Bonus point: Creatures that have been to space include flies, monkeys, frogs, turtles and spiders.

Test 4
German foods (20 points):
- Reis = rice
- Suppe = soup
- Zucker = sugar
- Brezel = pretzel
- Tischdecke = NOT A FOOD – it's a tablecloth

- Wassermelone = watermelon
- Mais = corn
- Salatgurke ('salad gherkin') = cucumber
- Honige = honey
- Mittagessen = NOT A FOOD – it's lunch
- Nudeln = noodles
- Schweinekotelett = pork cutlet
- Karotten = carrot
- Gabel = NOT A FOOD – it's a fork
- Becher = NOT A FOOD – it's a mug
- Krabbe = crab
- Auster = oyster
- Apfel = apple
- Speisekarte = NOT A FOOD – it's a menu
- Zitrone = lemon

Test 5

European words (10 points):
- passeggiata nello spazio (Italian) = spacewalk
- amarrage (French) = docking (similar to 'marriage')
- hjelm (Norwegian) = helmet (similar word)
- Fehlfunktion (German) = malfunction (similar word)
- vaisseau spatial (French) = spacecraft (should be able to get from 'vessel')
- brûler le moteur (French) = engine burn (could get 'burn' from 'crème brûlée')
- rifiuto spaziale (Italian) = space junk (could get 'rubbish' from its similarity to 'refuse')
- Luftsluse (German) = airlock (could get 'air' from 'Luft', Lufthansa, etc.)
- densità ossea (Italian) = bone density
- Sonnenkollektor (German) = solar panel (sun collector)

Bonus point: The languages spoken by my fellow astronauts included French, German, Italian and Norwegian.

Test 6
Italian sentence:
Italian: 'Ho riparato il pannello solare.' (1 point)
Translation: 'I repaired the solar panel.' (1 point)

Test 7
French sentences (5 points):
  1. Twelve astronauts have walked on the Moon.
  2. In space you float.
  3. You travel to space in a rocket.
  4. Astronauts repair the space station on spacewalks.
  5. The International Space Station is a collaboration between a
     number of countries.

Test 8
Chinese symbols (5 points):

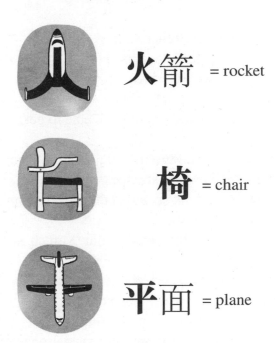

火箭 = rocket

椅 = chair

平面 = plane

 宇航员 = astronaut

 窗口 = window

Test 9
Match the meanings (10 points):
- **Group 1:** buoyancy = flottabilité (French), opdrift (Danish)
- **Group 2:** teamwork = lavoro di squadra (Italian), samenspel (Dutch)
- **Group 3:** training = entraînement (French), formazione (Italian)
- **Group 4:** parachute = Fallschirm (German), valskerm (Afrikaans)
- **Group 5:** constellation = Sternbild (German), stjörnumerki (Icelandic)

Test 10
English words (10 points):
- **ablation:** the wearing down of an outer surface over time, such as the erosion of a spacecraft's heat shield during re-entry
- **antipodal:** situated on the opposite side of the Earth
- **ephemeris:** a table of the positions of celestial objects, such as planets and stars
- **geostationary:** appearing to be stationary above a fixed point on the Earth
- **gimbal:** a device for keeping an instrument pointing in a certain direction, such as a solar panel towards the Sun

- **hypergolic**: the spontaneous ignition of a substance when mixed with another, particularly in a rocket
- **meridian**: a line on a spherical object running from North Pole to South Pole
- **perigee**: the point in an object's orbit when it is closest to the Earth
- **radiometer:** an instrument for detecting and measuring the intensity of electromagnetic radiation, such as visible light or radio waves
- **transponder:** a device that automatically transmits a secondary signal when it picks up a primary signal

TEST DEBRIEF

How did you score?

0–30

You don't find other languages easy to grapple with, which is not uncommon. Don't be disheartened – learning other languages takes practice, and with time almost anyone can learn another language. Perseverance is a great quality in an astronaut.

31–60

You have a good aptitude for learning other languages, and your English vocabulary is quite wide.

61–90

Excellent! You are a talented linguist and should be successful in learning other languages, if you put in the hard work. Well done.

HUMAN BEHAVIOUR AND PERFORMANCE (HBP) TRAINING: PART 1

So far in this book you have been examined on your 'hard skills', your intellectual capability and your language aptitude. However, the modern astronaut requires far more than this. Gone are the days when missions lasted just a few days or a week. Today's space travellers need to be effective crew members for at least six months aboard the ISS. During your stay you will share a small space with the same people – often from different cultures – day in and day out, in an unforgiving, remote environment. That requires exemplary interpersonal skills and the ability to understand human behaviour and how people react to various stressful situations.

We will explore all you need to know to think and act like an astronaut later in this section. First, try the following questions, to see how you would respond in different situations that are typical of life in orbit. Think about which underlying skills might be required.

TEST 1

While following a procedure for a science experiment, you realise that you have performed a couple of steps in the wrong order. You assess it likely that the end result will not be affected by this and that no harm has been done. Do you:

- a) Continue with the procedure, but make a written note to Mission Control about the incorrect order of the steps?
- b) Inform Mission Control of your mistake by voice, prior to continuing any further?
- c) Continue with the procedure, as you assess that it will have no impact?
- d) Ask a crewmate for a second opinion?

TEST 2

Another crew member is exercising on the cycle machine when you are scheduled to be using that equipment. Do you:

a) Go for a run on the treadmill instead?
b) Speak with Mission Control to reschedule your activities?
c) Interrupt the other crew member's exercise session to discuss the matter?
d) Get on with something else until the cycle machine is free?

TEST 3

You notice that a fellow crew member has done something in a way that isn't the most efficient way. Which two of the following four options are good practice?

a) Tell them in front of the group, so that everybody learns.
b) Ask them if they're interested in your feedback.
c) Ask them why they did it that way and explain why you would have done it differently.
d) Provide positive feedback alongside the negative.
e) Let them work it out for themselves – everyone has their own way of doing things.

TEST 4

You notice that there are not many coffee pouches remaining and there are still five days until you can open the next beverage container. You suspect one of your crewmates may have been drinking more than their fair share. Do you:

a) Take your own fair share of pouches and put them in your crew quarter, so that at least you won't run out?

b) Say nothing, as you can always open the next container early and avoid an awkward situation?

c) Have a quiet word with the person you suspect is drinking more than their coffee ration?

d) Explain to your crewmates that you all seem to be going through a lot of coffee, and do they have any suggestions about how to make the next container last long enough?

TEST 5

You are working in the airlock on one of the spacesuits. At the end of the task you notice that one of the tools is missing. What do you do?

a) Make a note to check the return air filters at the end of the day, because it will no doubt turn up there.

b) Tell your crewmates and ask them to keep an eye out for it.

c) Spend a few minutes searching for the tool.

d) Inform Mission Control that the tool is missing.

TEST 6

You notice one of the fans on the ISS has changed its rotation speed, suggesting that it may have developed a fault. Last time this happened you raised concerns, but were ignored. What do you do this time?

a) Nothing, as it is clearly not important or something would have been said previously.
b) Write it down, so that at least there's a record of it.
c) Send an email to Mission Control at the end of the day.
d) Raise it promptly with your fellow crew members.

TEST 7

One of your crewmates on the ISS has become withdrawn lately. Normally they are very talkative, but now they barely speak. Which two of the following could be useful actions?

a) Make a few light-hearted jokes and tell them to cheer up a bit – they should be happy, because after all space is a pretty cool place!
b) Ask them how they're doing and if everything is okay.
c) Give them a small token of support, by volunteering to take on an unpleasant task.
d) Report the situation to Mission Control – it could endanger the crew.

TEST 8

Your last phone call to your partner on Earth didn't go well. Things seemed stressed and the conversation ended badly. It has been weighing on your mind, and you notice that you're taking longer to do routine tasks this morning. Do you:

a) Work through it – everything will probably be fine tomorrow?
b) Confide in another crew member, just to let them know you're a bit preoccupied, if nothing else?
c) Speak with a trusted friend/relative/colleague on the ground who can help?
d) Ask Mission Control to add some breathing space into your schedule today?

TEST 9

You hear your commander giving an answer to Mission Control over space-to-ground radio that you think is incorrect. Do you:

a) Call down to ground the correct answer on the radio?
b) Assume the commander's answer is correct?
c) Speak with the commander face-to-face and raise your concern?
d) Speak to another crew member about your concern?

Answers

Compare your answers with those given below. In reality, it may not be possible to deal with the given situations as easily as the answers suggest. Astronauts are often task-oriented people who strive for maximum efficiency. However, tact, diplomacy and consideration for others are important traits for astronauts to possess. Here are the answers suggested by our experts.

Test 1

b) Inform Mission Control of your mistake by voice, prior to continuing any further.

When it comes to science experiments on board the ISS, you are probably not best qualified to assess the impact of your mistake. The experts on the ground will know better. Continuing with the procedure could make things worse. A quick radio call will clear up any ambiguity and prevent any loss of science data.

Test 2

c) Interrupt the other crew member's exercise session to discuss the matter.

Exercise is important in space. Your medical team on the ground is expecting you to do certain fitness training at certain times. Every daily activity in space is carefully coordinated against many different factors, so any changes to that plan should be discussed and the crew made aware of them. You may not want to interrupt your crewmate's exercise session, but a quick chat will clarify things immediately – they probably just overlooked the time!

Test 3

c) Ask them why they did it that way and explain why you would have done it differently.
d) Provide positive feedback alongside the negative.

There may be a good reason why your crewmate did it that way. It is best to first ask the logic behind their method and then tactfully suggest a different approach, if you think it could work better. You could ask if your crewmate is open to feedback, but in reality astronauts are continually receiving feedback on their performance, so the answer will be yes; and if it were no, that would raise a more difficult question: Are you going to give them feedback anyway? A good communicator will always try to point out both the positive and negative impacts of others' behaviour.

Test 4

d) Explain to your crewmates that you all seem to be going through a lot of coffee, and do they have any suggestions about how to make the next container last long enough.

This may seem like a trivial matter, but sometimes conflict can arise over the smallest things – like who drank all the coffee! With this situation, it is probably best to discuss this as a generic problem among the crew, without apportioning blame. Food and drink are carefully rationed in space and opening containers early is not an option, as it will only cause bigger problems later on.

Test 5

All of the above!

Losing a tool in the airlock is not something to stay quiet about. First, the airlock contains valuable and important equipment, such as the spacesuits. Second, the airlock needs to be a pristine location, so that when the crew open the hatch for a spacewalk they do not lose items into space that were left floating around. Mission Control can be really helpful in finding tools, and they can often spot items in camera views that the crew might miss. Other crew members need to be informed to keep an eye out, too, as the tool could turn up in a completely different location. The best chance of finding something is in the minutes immediately after it has gone missing, before it can drift too far. Finally, a more dedicated search may be required later on, in which case the airflow tends to move floating objects towards the return air filters . . . and so checking there, later in the day, could save a lot of time searching elsewhere.

Test 6

d) Raise it promptly with your fellow crew members.

Safety is paramount in space, so you should never be afraid to speak up, even if it's against someone more senior or someone who has discouraged your input in the past. A good crew member will always verbalise their safety concerns. It's always useful to discuss this as a crew first, so that the commander has all the information required, prior to making a call to the ground. For example, another crew member may have been working on the equipment recently.

Test 7

b) Ask them how they're doing and if everything is okay.
c) Give them a small token of support, by volunteering to take on an unpleasant task.

Part of being a good crew is looking out for each other. This can often mean checking how things are going, even when everything appears to be normal. After years of training together, crews will get quite used to each other's moods and will be able to tell when things aren't quite right. Light-hearted banter is all well and good, but you wouldn't want to use it in a situation where you think there may be a genuine problem, as it doesn't invoke a serious response. Reporting something to Mission Control, prior to talking to the crew member first, could betray their trust.

Test 8

Perhaps a bit of everything!

Part of astronaut training is designed to enable you to know yourself better. The sections on CAVES (see Photo Inset 1) and NEEMO (see Photo Inset 2) discuss this in more detail. It's important to know when you can work through a problem and when to speak up, especially if a problem is affecting your work. In this case, speaking to someone is probably the best option. Before flying to space, you will have identified a trusted network of friends and/ or relatives who can help you and your family in these situations. Other crew members can often help in these circumstances, too, and it is good for them to be aware if you're having a bad day. Finally, if you're falling behind in your work, it's always better to address this early with crewmates or Mission Control, rather than rushing to catch up and risk making a mistake.

Test 9

c) Speak with the commander face-to-face and raise your concern.

Everyone makes mistakes, so it's always good to question something if you believe it to be wrong. However, tact and diplomacy would suggest that you do this discreetly, face-to-face with your commander.

TEST DEBRIEF

How did you do? This section doesn't have a 'scoring' metric, as such; when it comes to complex human relationships, there is rarely – if ever – a right answer. However, you can see that there's a lot more to being an astronaut than simply whether you can operate machinery or fly a rocket.

To make sure that you are armed with the full repertoire of abilities needed for life in orbit, astronauts go through Human Behaviour and Performance (HBP) training. This encompasses all aspects of how humans interact, communicate, lead and follow orders in daily situations and, perhaps more importantly, during emergencies. Astronaut trainees are put in stressful situations and are made aware of their actions and communication skills afterwards. ESA's HBP course consists of five days of seminars on the subjects of communication, self-care, situational awareness, teamwork, decision-making, conflict management, observation, feedback and debriefing. During this time observable behaviours – known as 'behavioural markers' – are introduced.

Much of astronaut HBP training is based on exercises that were initially developed by the airline industry, although behavioural training practices from the business, medical and military worlds are thrown into the mix, too.

Trainees are put through their paces using three computer simulations known as Interlab, developed by the company Ninecubes Lernmedien, in cooperation with Lufthansa Flight Training. Participants must work in teams to coordinate their activities during a complex game. Trained behavioural observers provide feedback and also guide group debriefing sessions. At the end of the five days astronauts will have a much better understanding of their own personal behaviours, and of others' behaviours. This knowledge becomes vital for use in the next phase of HBP training – living for several days underground in a cave, as part of a multicultural group of astronauts.

In the next section of the book you will be able to try some problems and puzzles that will help you identify and test your own 'behavioural markers'. The goal is to build a toolkit of essential skills – known as 'competencies' by the HBP trainers. There is agreement among the international partners behind the ISS concerning which competencies and behavioural markers are crucial for crew members. The secret to success is to constantly analyse your previous performance and behaviour and feed that information into future activity, until the desired competencies become second nature. It is often best to play these games in groups, in order to test your team-working skills – so gather your friends, family or work colleagues to join in. However, it is also possible to do some of them individually.

COMMUNICATION

When communicating, one of the most important things to remember is to keep the message clear and concise. Wherever possible, include all pertinent information in the first message. This will prevent follow-up questions and excessive talking, which clogs radio channels and prevents other messages being sent. A good communicator also takes into account the person to whom they are talking. When talking with someone from a different profession, it is important to eliminate all unnecessary technical detail; you don't want to baffle them with a wall of jargon. On a multinational project such as the International Space Station – with astronauts and mission controllers from many countries around the world – astronauts must avoid using slang, idioms, sayings and cultural references that might not translate to non-native speakers. And, of course, sending a clear and concise message is only half the battle. The person receiving the message has a responsibility to listen carefully and interpret it correctly.

It's surprising how many accidents and mistakes have occurred due to poor communication. When I was training to be an astronaut, the Tenerife airport disaster was one of the first things we learnt about, as part of the astronaut class of 2009. In 1977, two Boeing 747s ('Jumbo Jets') crashed into each other on the runway of the Spanish holiday island. A total of 583 people were killed, making it the deadliest accident in aviation history. KLM Flight 4805 had tried to take off while Pan Am Flight 1736 was still on the runway. Foggy conditions severely hampered the situation. We spent time discussing the actions of the various parties involved in the accident, and how it could have been avoided.

Below is part of the transcript of communication between the control tower and the aircraft involved. The Pan Am crew were struggling to identify the correct runway exit in thick fog, while the KLM captain was in a hurry to get airborne:

Time	Source	Message
1702:03.6	Pan Am Co-pilot	Ah – We were instructed to contact you and also to taxi down the runway, is that correct?
1702:08.4	TENERIFE TOWER	Affirmative, taxi into the runway and – ah, leave the runway third, third to your left.
1702:16.4	Pan Am Co-pilot	Third to the left, OK.
1702:18.4	Pan Am FLT ENGR	Third, he said.
1702:20.6	TENERIFE TOWER	Third one to your left.
1702:21.9	Pan Am CAPTAIN	I think he said first.
1702:26.4	Pan Am CO-PILOT	I'll ask him again.
1703:12.1	Pan Am CO-PILOT	Must be the third ... I'll ask him again.
1703:14.2	Pan Am CAPTAIN	OK.
1703:29.3	Pan Am Radio(c/p)	Would you confirm that you want the Clipper one seven three six to turn left at the third intersection?
1703:35.1	Pan Am CAPTAIN	One, two.
1703:36.4	TENERIFE TOWER	The third one, sir; one, two, three, third, third one.
1703:40.1	Pan Am CAPTAIN	That's what we need, right, the third one.
1703:42.9	Pan Am FLT ENGR	Uno, dos, tres.
1703:44.0	Pan Am CAPTAIN	Uno, dos, tres.
1703:44.9	Pan Am FLT ENGR	Tres – uh – sí.

And a few minutes later:

1705:44.8	KLM (Radio)	Uh, the KLM four eight zero five is now ready for take-off ... uh, and we're waiting for our ATC clearance.
1705:53.4	TENERIFE TOWER	KLM eight seven zero five, uh, you are cleared to the Papa Beacon, climb to and maintain flight level nine zero, right turn after take-off, proceed with heading zero four zero until intercepting the three two five radial from Las Palmas VOR.
1706:09.6	KLM (Radio)	Ah roger, sir, we're cleared to the Papa Beacon flight level nine zero, right turn out zero four zero until intercepting the three two five and we're now (at take-off).
1706:13.0	KLM CAPTAIN	We're going.
1706:18.19	TENERIFE TOWER	OK.
1706:19.3	Pan Am Radio(c/p)	No ... eh.
1706:20.08	TENERIFE TOWER	Stand by for take-off, I will call you. (message not heard by KLM crew)
1706:20.3	Pan Am Radio(c/p)	And we're still taxiing down the runway, the Clipper one seven three six. (message not heard by KLM crew)
1706:25.6	TENERIFE TOWER	Roger, alpha one seven three six, report when runway clear.
1706:29.6	Pan Am Radio(c/p)	OK, we'll report when we're clear.

1706:32.43	KLM FLT ENGR	Is he not clear then?
1706:34.1	KLM CAPTAIN	What do you say?
1706:34.15	KLM-?	Yup.
1706:34.7	KLM FLT ENGR	Is he not clear, that Pan American?
1706:35.7	KLM CAPTAIN	Oh yes – emphatic.
1706:50.0		Collision.

The full circumstances that caused that tragic accident were complex, but it's easy to see how poor communication, misunderstanding and confusion played a major part in the collision.

The following tests in this section are variously designed to test individual and group communication situations. Communication is the secret to a well-oiled team and is a vital skill for you to master on your path to space. Good luck!

TEST 1: Phonetic alphabet

The NATO Phonetic Alphabet (A for Alpha, B for Bravo, etc.) is often used by astronauts to cut down the risk of a verbal misunderstanding. Astronauts will be expected to memorise this phonetic alphabet during their training. The full phonetic-alphabet chart is outlined at the end of this section. How many of the corresponding letters and words can you complete in the chart below, without looking at the full chart? *Note:* If you do not know any of the words to start with, study the chart for five minutes and then try to remember as many of the words as possible. Keep repeating the test until you can complete the whole chart.

Letter	Word
A	Alfa/Alpha
B	Bravo
C	
D	
E	
F	
G	
H	
I	
J	
K	
L	
M	
N	
O	
P	
Q	
R	
S	
T	
U	
V	
W	
X	
Y	
Z	

TEST 2: Picture sequence – group exercise

In space there are often times when you need to communicate with someone else about something that you can see, but they can't, such as when you are repairing equipment on a spacewalk and you are talking to Mission Control. It is important that this information is relayed quickly and accurately. An effective team will be able to do this very well.

This test works best with a group of four or five people – an ideal family activity – and all you need is a pack of playing cards and a timer. To begin, remove any extra cards (such as jokers) to leave a standard 52-card deck. Shuffle the deck and deal two cards face-down to each member of the team. Do not show your cards to anyone else.

Your team's goal is to arrange the cards in a row in numerical order (ascending), face-down, without showing the cards to each other. You can only communicate verbally with each other.

Start the timer. Once you have successfully worked together to figure out the order (again, without showing each other the cards), team members should lay their cards face-down in a row and stop the timer. Turn over the cards, to see if the order is correct. How long did you take?

Repeat the test. Can you improve your time by honing your communication skills? Once you have got used to one sequence, try the other sequences below. Do different communication tactics work better, depending on the exact sequence that you are attempting to put the cards in?

- Numerical order (descending)
- Alphabetical order (e.g. Four of Hearts before Four of Spades, before King of Clubs, etc.)
- All blacks, followed by all reds *
- Odd numbers, then even numbers *
- Hearts, then Clubs, then Diamonds, then Spades *

(* The numerical order doesn't matter.)

TEST 3: Word sequence

Perfecting the art of working together to put playing cards in a given sequence is one thing, but what if you're not told what the sequence is?

In this task for teams of two to four people, your team is given a list of four words. You must write the words on four separate pieces of paper and distribute them as evenly as possible among your team members.

Alongside your list of four words you will be given two more words, placed at the beginning and end of your other words, to form a six-word sequence. As a team, your challenge is to work out how to put the four words in the correct order between them.

Before you begin, here is an example. The four words you are given are: car, bike, plane, train. You then have to put them in order between: foot . . , . . . , . . . , . . . , rocket. In this example, the sequence is the slowest to fastest modes of transport, so the correct order would be: foot, bike, car, train, plane, rocket.

This task is about critical thinking and collaboration. Sometimes the answer will be obvious, but not always. Can you arrive quickly at the right answer by listening carefully to everybody else's ideas? To help you, the team is allowed three Internet searches. The team must decide collectively when it is best to use these opportunities.

Below are ten lists of four-word sequences. Beneath are the six-word sequences they must be added to, in the correct order.

1. Spider, Dog, Wasp, Human
2. Lima, London, Cairo, Beijing
3. Humerus, Clavicle, Fibula, Patella
4. Goblet, Order, Prisoner, Chamber
5. Pink, Brown, Blue, Green
6. Neptune, Mars, Saturn, Jupiter
7. Helen, Abigail, Nina, Lucie
8. Four, Five, Three, Four
9. Croatia, France, Canada, Mozambique
10. Tetchy, Joyful, Foolish, Medic

1. Snake, _____, _____, _____, _____, Crab

2. Oslo, _____, _____, _____, _____, Canberra

3. Cranium, _____, _____, _____, _____, Metatarsals

4. Philosopher's, _____, _____, _____, _____, Half-Blood

5. Yellow, _____, _____, _____, _____, Black

6. Earth, _____, _____, _____, _____, Uranus

7. Sarah, _____, _____, _____, _____, Erica

8. Two, _____, _____, _____, _____, Four

9. Chad, _____, _____, _____, _____, Mauritania

10. Shy, _____, _____, _____, _____, Tired

TEST 4: Non-verbal communication

A lot of human communication is non-verbal. Our facial expressions sometimes give away how we are feeling without the need to say anything, and a quick hand-signal can often convey a message in a fraction of the time it would take to express it verbally. This can be a useful form of communication in conditions where verbal communication is either impossible or undesired. However, these signals can often be misconstrued, and if a non-verbal message is to be conveyed clearly, it is important that both sender and receiver are able to convey and recognise the correct emotion.

This exercise explores the ability to express emotion and information through body language. First, take a look at the following photographs and write down the emotion that best fits the facial expression.

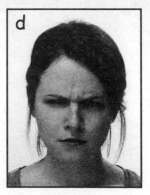

Next, having tested your emotion recognition skills, see if you can accurately convey the correct non-verbal information. With a partner or multiple group members, take it in turns to write an emotion on a piece of paper, before folding the paper over and adding it to a pile. Try and get a mix of emotions, beyond those in the images above. Try: ashamed, confused, jealous, lonely, relaxed, excited, stressed.

Once there are around a dozen bits of paper, you and your partner or group should take it in turns to draw one from the pile. Your task is to act out that emotion using only your facial expressions – your partner or the rest of the group has to guess what word was on the paper. Take it in turns to act out the emotions. If you are playing on your own, shuffle the pieces of paper, pick one at random and then make the face in front of a mirror, as quickly as you can. Do you think you are convincing?

Now try including some messages. Do the same thing again, but this time include hand-signals and other body language to help convey the message. Here are some messages you may wish to try: wait there, that's correct, be careful, proceed, I didn't hear you, do that again, watch me, I can't see, emergency stop, I'm not sure.

Once you have completed these exercises, have a discussion within the group, or reflect on it if you are by yourself. Were some emotions or messages easier to convey than others? Were there some that were easily mistaken for another? Did you improve as you became more aware of how your own expressions or body language were being interpreted by others? And did you get better at recognising emotion in others? How might this knowledge be helpful for life in orbit?

Answers

Test 1

Letter	Word
A	Alfa/Alpha
B	Bravo
C	Charlie
D	Delta
E	Echo
F	Foxtrot
G	Golf
H	Hotel
I	India
J	Juliet
K	Kilo
L	Lima
M	Mike
N	November
O	Oscar
P	Papa
Q	Quebec
R	Romeo
S	Sierra
T	Tango
U	Uniform
V	Victor
W	Whiskey
X	X-Ray
Y	Yankee
Z	Zulu

Test 2

Discuss with your teammates how the test went, and try to come up with ways to play the game more efficiently. Is there a way to decide who should speak first, for example? Shuffle the pack and deal two more cards to each player and try again. Can you beat your previous time?

Ultimately, it is not important which sequence you use. The key thing is that the team has a shared goal and that each team member has unique information that is crucial to achieving it. By discussing your performance you are creating a shared mental model that enables you to communicate efficiently and effectively, as well as identifying ways of improving next time.

Test 3

1. Snake, Human, Dog, Wasp, Spider, Crab (0, 2, 4, 6, 8, 10 legs)
2. Oslo, London, Beijing, Cairo, Lima, Canberra (North Pole to South Pole)
3. Cranium, Clavicle, Humerus, Patella, Fibula, Metatarsals (bones: head to toe)
4. Philosopher's, Chamber, Prisoner, Goblet, Order, Half-Blood (word after 'The' in successive Harry Potter books)
5. Yellow, Green, Brown, Blue, Pink, Black (snooker balls in ascending order of value)
6. Earth, Jupiter, Mars, Neptune, Saturn, Uranus (alphabetical order)
7. Sarah, Helen, Nina, Abigail, Lucie, Erica (each name starts with the last letter of the one before)
8. Two, Three, Five, Four, Four, Four (each term is the number of letters in the previous word)
9. Chad, France, Canada, Croatia, Mozambique, Mauritania (number of vowels = 1, 2, 3, 4, 5, 6)
10. Shy, Medic, Foolish, Tetchy, Joyful, Tired (synonyms for the first six of Snow White's dwarfs in alphabetical order = Bashful, Doc, Dopey, Grumpy, Happy and Sleepy)

Test 4

It can be difficult to recognise facial expressions and determine the correct emotion or body language. Some people are much better than others both at sending clear non-verbal signals and at interpreting them. However, with practice you can improve at this. Here are the emotions that the pictures were trying to convey:

a) Happiness
b) Fear
c) Embarrassment
d) Anger
e) Disgust

TEST DEBRIEF

How did you find this section? Did you find the individual tests easier than the group tests? Did your group have arguments or get frustrated with each other?

One of the key skills for any astronaut is the ability to communicate, whether it's with other astronauts, Mission Control or the public, once they've returned from orbit. The astronaut selection process deliberately searches for people who naturally have good communication skills, but HBP training focuses heavily on developing astronauts and equipping them with the communication tools they will need in order to become a successful space traveller and an effective crew member.

Communicating information clearly and concisely as an astronaut means using common and understandable words. But in some situations astronauts must also use very precise, technical language, all the while relaying a message using as few words as possible. Here's a short transcript from part of my spacewalk. Some of it will be hard to understand without first knowing the technical abbreviations, and yet it would be clearly understood by ground and space-station crews.

Ground: 'Can you catch us up?'

Me: 'Sure, I've got the Node 1 cable and I've just put a wire tie on Lab 023 forward handrail and temp stowed the MDM coil.'

Ground: 'Alright, that's great news. We have a caution to avoid contact with Cygnus thrusters and the CBM petals while you're back in this area and then you'll be heading back to mate the V605-J605 connection on Node 1.'

In space no one likes surprises, so when time permits, you should always communicate your intention before taking action. This enables others to prepare and react or help, if required. It also allows others to raise concerns and anticipate any problems that your actions might cause. I remember during my mission that, about three hours into my spacewalk, while the ISS was over Oklahoma and Arkansas, I was moving back down towards the Quest airlock to stow some equipment when I noticed that my tether was in danger of getting tangled between two cameras. So I told my spacewalk colleague – NASA astronaut Tim Kopra – about the problem and the action I was going to take to sort it out.

Clearly the ability to communicate and work effectively as a group is a vital skill for astronauts, especially during complex emergency scenarios on the ISS, where the commander may have to delegate many important tasks and closely coordinate this activity. If certain procedures are completed out of synch, it could make the difference between mission success or losing complete control of the space station. To avoid such a scenario, below are some of the key tenets for any trainee astronaut to consider, when it comes to communication.

ASTRONAUT ESSENTIALS: COMMUNICATION

- Always ask questions before the need for the information is urgent, allowing you to anticipate problems before they escalate.
- Communication is a two-way process. A good communicator will optimise conditions to ensure each message is delivered coherently. A good listener will pay maximum attention to ensure all information is understood.
- Confirm that what you've said has been understood correctly, and explain it again using alternative language if it hasn't.
- Being a good communicator is often about choosing when to speak and when to act first.
- During important conversations, maintain an attentive posture and eye contact so that the other person knows you are listening.
- Wait until the other person has finished talking before responding.
- Make sure quiet members of the group are asked about their thoughts and reactions.
- Thank people for their participation and treat opinions, feedback and suggestions with respect.

MAYHEM IN THE METROPOLIS

I want to finish this section by examining a final
communication test that is part of the second
phase of the astronaut selection process.

    The test is as follows. You have to work closely
with another candidate on a computer-based task. You can only
communicate with each other over a voice-loop – you can't see the
other candidate or what's on their screen. All the time you're both
being watched by a psychologist.

    Between you, you need to manage the flow of traffic at a series
of busy junctions in the heart of a major city. You control half the
traffic lights, while your partner controls the other half. Together
you must keep all the vehicles moving and prevent traffic jams.
However, during the exercise each candidate will receive individual
goals to achieve on their screen. It could be that you need to let
more blue cars through the junction than red. Equally, your partner
could receive the opposite target of letting mostly red cars pass.
Or it could be something else entirely. It is up to you whether or
not you communicate your individual instructions to your partner.
Think about what you would do, before reading on.

It is not hard to see why this test formed part of the selection process. At a basic level, it tests your calmness under pressure and your ability to communicate with someone else quickly and accurately. It is also a good test of your spatial-awareness skills, as you are moving objects through a limited area in the most efficient way possible. Beyond that, it tests your teamwork skills and your inclination to collaborate. Did you decide to try and work with your partner to achieve both your individual goals and your joint mission?

Some people really struggled with this test, emerging from the room visibly distressed. It was clear that in some cases there had been a complete breakdown in communication – some candidates may have valued their individual score more highly than working together to achieve the team goal. Those who were more relaxed generally fared better, particularly if they approached the task with a light-hearted touch and a sense of humour. Bear this in mind as you progress now in your training to more team-based tasks.

TEAMWORK AND GROUP LIVING

The International Space Station is a challenging environment. As an astronaut, you have to live in the unusual setting of weightlessness with the same people for months at a time. Your crewmates will have different professional and cultural backgrounds, varying ways of doing things and individual emotional needs. How will you ensure that you live alongside each other successfully and pull together as a team to get the job done?

One of the key skills is collaboration. In space, you will often work on individual tasks during the day, and keeping up with a tight timeline can be tough. Rather than seeing this as a competition with your crewmates, it is vital to maintain a team perspective. If one person falls behind the timeline, it's a problem for the whole team, and that includes Mission Control. Collaboration is vital – you are a unit, all pulling together to achieve one thing: mission success.

It is also important that astronauts take responsibility for their own actions and mistakes. There is no problem in celebrating successes as a team, but you should also be ready to admit your own failures. Allowing a mistake to go unreported could cause a safety issue, or at the very least enable the same mistake to occur again in the future. If you were the last person to use a piece of equipment in space that developed a fault, you are expected to say so.

Delivering an outcome for the team has priority over actions that will give you personal credit or a sense of individual professional achievement. Good team members demonstrate patience, respect and appreciation of each other. You should also consider how to effectively distribute and accomplish work, according to the skills and abilities of the group. Take the culture and personality of other crew members into account, and understand the strengths and weaknesses of yourself and of fellow team members. Be tolerant of difficulties that arise from these differences.

The following group and individual tests will examine your team-work skills through a variety of methods. By putting your team's goals ahead of your own, you will be one step closer to space.

TEST 1: The arm exercise

This is a game for two players. If you are in a larger group, you can do this exercise in teams, and pair up one member of each team with a member of another team.

The two players lock hands like this, with their elbows resting on a table.

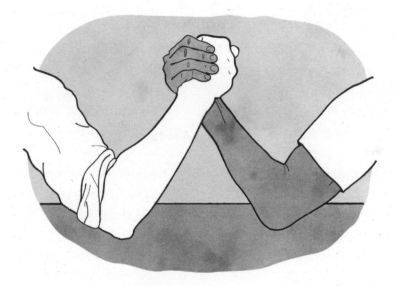

Each round of the game lasts for 30 seconds and there are only two rules:

- You win 1 point every time the back of your partner's hand touches the table.
- Your goal is to get as many points for yourself as possible. Your partner's performance is irrelevant.

Try this exercise several times. If more than two people are playing, compare everyone's scores afterwards. Who did best? Why? See the Answers below.

TEST 2: I'm an astronaut . . . get me out of here!

The following is an example from one of the group tests that formed part of the astronaut selection process for the class of 2009. Six potential candidates were all working together. How would you approach the task, if you were one of them?

You are part of a group of six explorers needing to return to base camp on foot, after your mission has been aborted. However, your communication devices have failed. You have high mountains to the west and dense rainforest to the east. To the north is a hostile encampment, and the lake to the south of you is crocodile-infested. You do not have any weapons, you possess minimal equipment and only enough food and water for one day. How do you get back to base camp?

Each candidate is given just a few minutes to read through the task and start developing initial ideas. The group then comes together to discuss it and work out a solution. After half an hour one member of the group has to give a two-minute presentation on your collective plan. How would you set about solving this problem?

Even better, get five friends or family members to join you and try to tackle this exercise as a group, before reading the comments in the Answers.

TEST 3: Ranking exercise

Astronauts undertake a number of exercises to demonstrate the importance of being part of a team, and how input from different people with varying perspectives is incredibly valuable. Here is a simple way to discover these things for yourself.

Your job is to rank a list of ten items. Team members initially do this individually, with no discussion allowed. You have five minutes, and you will get 1 point for every item in the correct position.

In the second half of the task, team members are allowed to discuss the ranking with the rest of the team. After ten minutes the team must reach a consensus concerning the order of the items. One point is still awarded for each word in the correct position.

After both halves, the correct ranking is revealed. How do your individual scores compare with the team score? Can you use what you've learnt from the team situation to improve both the team score and your individual score in the next round?

Round 1: Common English words

The compilers of the *Oxford English Dictionary* have scoured texts from across the English-speaking world to create a list of 2.1 billion words, and how often they appear in the English language. Their research found that the top ten most common words are: *of*, *be*, *and*, *a*, *that*, *have*, *the*, *to*, *in* and *I*. Rank them 1–10, from most common to least common.

Round 2: Highest-grossing films

When you take into account worldwide box-office takings, these are the highest-grossing films in history. Can you put them in the correct order, starting with the highest-grossing? *The Avengers*, *Black Panther*, *Titanic*, *Jurassic World*, *Star Wars: Episode VIII – The Last Jedi*, *Avengers: Age of Ultron*, *Avatar*, *Furious 7*, *Harry Potter and the Deathly Hallows: Part 2*, *Star Wars: Episode VII – The Force Awakens*.

Round 3: Hopes and dreams

In 2016 the Nationwide Building Society asked 2000 people of all ages across the UK about their dreams for the future. Here are the top ten most popular answers, which you need to rank: owning a dream car, renovating a home, getting in shape, visiting a dream holiday destination, retiring early, owning a home, getting a dream job, seeing one of the Seven Wonders of the World, learning a new skill (music, language, etc.), travelling around the world.

TEST 4: Creative collaboration

With the crew of the ISS changing regularly, you often pick up where somebody else has left off. You have to be able to continue somebody else's work with minimal disruption. Put your collaborative skills to the test with this group drawing exercise. You need only a pen, some paper and a timer.

One person in the team starts by drawing a shape or outline. After five seconds the drawing is then passed to the next team member, who must add to it and pass it on, and so on. You are not allowed to discuss what is being drawn, either before or during the exercise. The task stops after one minute and is only successful if there is a completed drawing of something recognisable.

At the end of the task your team should debrief on the test. See the Answers for more tips on this.

Answers

Test 1

This simple exercise is a great example of how our preconceptions can lead us astray, particularly if we always assume that a competition has to be adversarial. An arm wrestle is such a familiar exercise that you may have made the mistake of thinking that this task was indeed an arm wrestle. So you may have spent your 30 seconds straining to beat your 'opponent'. However, if you collaborated with each other as partners, you may have realised the trick to gaining more points. You simply work together to rock your arms back and forth, alternately touching the backs of your hands on the table. Remember, you only care about getting as many points for yourself as possible – there is nothing in the rules that says you have to get more points than your partner. Often communication and collaboration are surer paths to success than direct competition.

Test 2

What was the point of this task? 'It was amazing to see how differently the groups worked,' says Gerhard Thiele, ESA astronaut and head of ESA's astronaut selection process in 2008. 'The key wasn't the difficulty of the task, but seeing how people reacted.' He personally saw 100 candidates go through this exercise. Some people used half their allotted time to work out who should give the presentation at the end, instead of working on the problem.

Others immediately tried to stand out by adopting the leadership role, but ignored other people's input, unless it agreed with their own ideas. 'That is obviously something you don't want to see in an astronaut, at the end,' says Thiele. Such behaviour immediately put a question mark next to their application. However, one action was not enough to kick someone out. If a second test revealed the same behaviour, that would see them eliminated from the running. Strong candidates facilitated equal discussion within the group, actively including people whose voices weren't naturally being heard and ensuring the maximum number of ideas in the mix.

The underlying secret was that there was no good answer to the problem. With the explorers hemmed in on all sides, this was an impossible task. Instead the exercise was designed to help selectors assess the qualities of the candidates' approaches in a group setting – to identify the key skills of leadership and followership.

In some ways it doesn't matter what the actual question is; it is more about seeing how you respond to adversity and how you handle collaboration. In another question, the team were faced with a boat coming into a harbour to drop off cargo containers onto lorries, which would then take them to the train station. Candidates had to contend with so many variables – rising tide levels, train timetables, maximum lorry loads, and so on – that it wasn't feasible to work out the perfect solution in just 30 minutes. It was about how they made the best of the situation, as a member of a team.

Test 3

Words: *The, be, to, of, and, a, in, that, have, I.*

Films: *Avatar, Titanic, Star Wars: Episode VII – The Force Awakens, Jurassic World, The Avengers, Furious 7, Avengers: Age of Ultron, Harry Potter and the Deathly Hallows: Part 2, Black Panther, Star Wars: Episode VIII – The Last Jedi.*

Dreams: Travelling around the world, visiting a dream holiday destination, owning a home, getting in shape, seeing one of the Seven Wonders of the World, owning a dream car, retiring early, learning a new skill, renovating a home, getting a dream job.

This test often highlights the benefits of good collaboration. As an individual, it's extremely hard to put these in the correct order, since the answers are based on a group consensus, which is unlikely to match precisely your personal viewpoint. However, good teamwork, a constructive discussion and sharing of ideas will often result in the team scoring more highly than any one individual was able to achieve.

Test 4

In your team's debrief to this exercise, you should discuss what worked well and what didn't. Did people tend to panic under the time pressure? Did they manage to give the next person enough to go on, in order to continue the drawing? Discuss within the group some tactics for improving the team's performance in future (remember that you still can't discuss precisely what you're going to draw).

TEST DEBRIEF

Today astronauts are selected from a diverse range of backgrounds, careers, cultures and personal experience. This broadens the skill-set and potential of the crew as a whole, but also increases the necessity for good teamwork and collaboration. In order to embrace this diversity and maximise its strengths, astronauts will often have to display a delicate balance of leadership and followership.

Sometimes you may need to be forceful enough to be heard, but not so headstrong and inflexible that you cannot accommodate others' opinions and assess their benefits. Consideration of others is vital if a crew are to be successful, whether they are living together for a few days in a cave or for several months in space. And of course, as with any good team, there is a need for discipline and respect.

During astronaut training there is a constant requirement for good teamwork, and plenty of opportunity to develop these skills – be it planning an EVA, winter survival training in Russia, operating the Soyuz spacecraft or dealing with emergencies on the ISS.

ASTRONAUT ESSENTIALS: TEAMWORK AND GROUP LIVING

- Always use 'I' rather than 'us' or 'we' when talking about your errors.
- Never be reluctant to ask a team member for help, rather than trying to deal with something on your own – even if your actions led to the issue in the first place.
- Avoid infringing upon other team members' responsibilities and task allocations, or 'back-seat driving' by telling someone else how you think they should do their job.
- There are different ways of working and 'different' doesn't mean better or worse. Respect and appreciate these differences without trying to change them.
- Always put collective mission goals before your own personal targets.
- Invite colleagues to join team activities, especially when they are shy to participate.
- Let your actions show others that when you commit to do something you will get the job done to a high standard.
- Be aware of how your own living habits may be affecting others, and be prepared to adapt them accordingly.

SITUATIONAL AWARENESS

Good situational awareness is important in many aspects of life. Being aware of your surroundings will not only keep you and your colleagues safer, but will also help you to make more informed decisions and, ultimately, will allow you to enjoy living more fully in the present.

Astronauts on the ISS need to pay attention to their surroundings at all times. Situational awareness relies on using all the senses available to you and employing this information to create an accurate mental model of your surroundings. Interestingly, although you may think your eyesight dominates this mental model, your other senses can be equally important. For example, your brain processes sound up to 100 times faster than it does vision, and background noise can give you vital clues to a dangerous situation developing. Particularly important on board a space station is your sense of smell. Fire is one of the most dangerous emergencies that astronauts can face. A good sense of smell can detect a change in materials warming up, even before they start smouldering and ignite.

Astronauts are trained consciously to look out for and listen for things. Whenever I looked out of the Cupola window (the large viewing window on the ISS) I would always spend a few moments checking the outside of the Soyuz and the ISS's structure, for any signs of damage from space junk or micrometeorites. It was on one of these occasions that I identified a leak coming from one of the tanks on the Progress resupply vehicle. Spend enough time in space and you can become really good at being situationally aware. Scott Kelly, a NASA astronaut who spent a year on the ISS, was so in tune with the way his body was feeling each day that he knew when the carbon-dioxide levels had changed, even before he looked at the readings.

Perceptive astronauts will constantly be monitoring the people, systems and environment around them and will be alert to changes. Such situational awareness is part and parcel of your commitment to the safety of operations and to mission success.

TEST 1: What's missing?

Here is an exercise to help improve your situational awareness. You will need a watch or timer, set to ten seconds.

1. Without looking at it just yet, note that below is an illustration of a tray with 15 random household items of various sizes, shapes and patterns. When you are ready, start your timer for ten seconds. Your task is to look at the diagram for ten seconds only, then try and remember as many of the items as possible, including their positions and patterns. Draw or write as many of the items as you can in the correct positions, and with the correct patterns, on the corresponding tray on page 141.

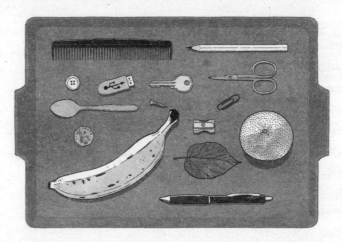

Note: You can play a variation of this game at home. All you will need is a tray, a cloth to cover it and 15 random items. One player/team places all 15 of their items on the tray, in a random order. They then show their opponents the tray for ten seconds, before covering it with the cloth. How many items can the other team remember?

2. Without looking at it just yet, note that on the next page is an illustration of a tray with 13 random items. When you are ready, start your timer again for ten seconds. You have ten seconds to remember all of the objects on the tray, but this time the

illustration of the tray on page 142 will have a missing object. Your task is to say which item has been removed. To complicate matters, the position of the objects has been changed in the second illustration.

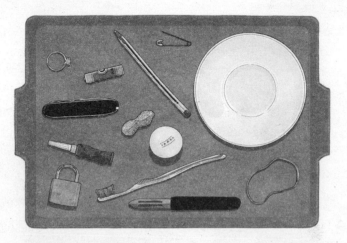

3. Again without looking at it just yet, note that below is an illustration of a tray with ten random items. When you are ready, start your timer for ten seconds once again. You have ten seconds to remember all of the objects on the tray, but this time some of the objects on page 142 will have different patterns. They have also been rearranged differently on the tray. Your task is to say which items have different patterns.

TEST 2: What happened?

STOP: BEFORE YOU READ ON, STUDY THE PICTURE ABOVE FOR 60 SECONDS. THEN, COVER THE PICTURE WITH YOUR HAND, AND AND SEE IF YOU CAN ANSWER THE FOLLOWING QUESTIONS:

1. How many cyclists did you see?
2. How many people are in the picture?
3. How many cars were damaged in this accident?
4. What object was lying on the ground?
5. What injury did the man on the ground seem to be suffering from?
6. What was the bus number?

TEST 3: The straw and the toothpick

LOOKING STRAIGHT AHEAD

PERIPHERAL VISION

People who have good situational awareness skills are able to keep track of things they are not directly looking at. This peripheral vision is one of the first things to go when experiencing high g-forces in a centrifuge or rocket. It is all about being optimally aware of your environment – if your peripheral vision starts to wane it is a sign you may be in trouble. During training in Star City, astronauts experience an acceleration equivalent to eight times their body weight (8g) in a centrifuge for 30 seconds. During this time, a number of tests are conducted, one of which is to focus straight ahead and read out numbers on a display while clicking a button every time you see a small light flash in your peripheral vision.

You may not be able to use a centrifuge, but you can still test out your peripheral vision skills using a straw, a cup and a toothpick/cocktail stick.

Place the straw in the cup. While looking straight ahead at an object in front of you, move the cup as far to the left or right as you can until it is at the very edge of your peripheral vision. While continuing to look straight ahead, can you put the toothpick in the end of the straw? If you can't, bring the cup closer to the centre of your vision until you can. Then start working your way back out again.

TEST 4: Assemble a puzzle as a team under time pressure

Astronauts sometimes need to repair large objects in space, such as satellites and solar arrays. This is sometimes carried out during a spacewalk. Other times, objects have to be put together quickly and accurately, such as when the crew of Apollo 13 had to improvise a way to install filters to remove dangerous levels of carbon dioxide inside the spacecraft.

When assembling or maintaining objects, astronauts must have good dexterity, hand–eye coordination and work as a team. They must also be able to manipulate tools and objects while wearing a pressurised spacesuit that includes gloves over their hands. This is exactly what we practise during our Neutral Buoyancy Training, which is training conducted in swimming pools to re-create some of the conditions of being on a spacewalk, as the neutral buoyancy of water provides conditions somewhat similar to weightlessness.

The gloves, designed to give protection from the space environment, can be thick and bulky, but are made so astronauts can move their fingers as easily as possible while on an EVA. A piece called a bearing connects the glove to the sleeve, allowing the wrist to turn. Astronauts must learn to work with their gloves on, to handle both large and small objects. Re-create this challenge by attempting the following team task.

You will need:

- A jigsaw puzzle with a small number of pieces (25 is ideal)
- Two pairs of gloves for each team member – one pair of normal woollen gloves, the other a pair of gardening gloves
- A stopwatch

Each team member puts on two pairs of gloves – first the woollen ones, then the gardening gloves over the top. Time yourselves on the stopwatch to see how quickly you can work together as a team to assemble the puzzle. After your first attempt, have a discussion with your crew about what worked well and what didn't. Come up with a new approach to the problem and try again. Did you beat your first time? Keep debriefing with your teammates after each attempt and see how much you can improve your time by using effective communication and problem-solving skills.

Switch it up. This and the other team tasks in this book can be done by two teams competing against each other. Can you improve enough to beat your opponents? A great way to test your skills is by switching the teams halfway through. Can you adapt to working with the people you were just competing against? Can you quickly communicate the tactics you picked up by working with the previous team? This situation is very similar to one experienced by the six ESA astronauts selected in 2008 during our HBP training. We were competing in two teams of three to solve a challenge related to the exploration of Mars. However, the trainers mixed the teams up part-way through the exercise and we had to adapt.

Answers

Test 1

Note: If you played this game with a real tray in teams at home, this exercise can be repeated many times. If teams become too familiar with the items, simply swap them out for new ones. If you become proficient with 15 items, try increasing it to 20 or 25. As always, the key is to debrief with your teammates after each round, to discuss how it went and whether there is a better system that will enable you all to work together more efficiently. Can you consistently improve your performance?

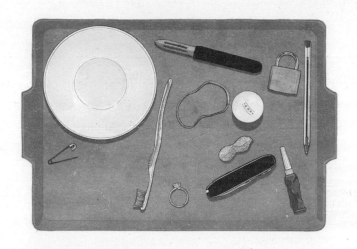

Test 2

1. Two.
2. Eight.
3. Two.
4. A car wheel or tyre.
5. A leg injury.
6. 249.

ASTRONAUT ESSENTIALS: SITUATIONAL AWARENESS

- Raising an issue early allows you to address it before it becomes a bigger problem.
- Monitor yourself and others for signs of stress, fatigue, complacency and task saturation.
- You can help others by not interrupting them during important procedures.
- Give particular focus to information that could affect safety and other critical tasks.
- Trust your gut instinct and always speak up.
- If someone else voices a concern, it is important to acknowledge and consider it, and not to simply discard it out of hand if you disagree.
- Never switch off, even during periods of low activity. Instead you should remain aware of your surroundings in order to pick up abnormal noises or other subtle indicators of potential issues.
- When conflicting information contributes to a problem, you should always check the information from each source.

HBP TRAINING: PART 2

LOGIC AND REASONING

Astronauts are often faced with situations where there are a number of variables with multiple possibilities, and finding the correct answer may seem a daunting task. An example of this is planning a spacewalk. Imagine trying to work out what tasks need to be done on a spacewalk, in which order and by which crew member. Some of the factors to bear in mind might be:

- What are the best routes to take (you need to be mindful of crossing paths with another spacewalker, due to the risk of tethers becoming tangled)?
- What tools are required, and will they all fit in the toolbag?
- What equipment will fit in the airlock along with two spacewalkers?
- Will you work together or as individuals on each task?
- Are there restrictions on doing some tasks in daytime or night-time (every 45 minutes there is a sunrise or sunset)?
- Are there any areas that are dangerous or out of bounds?
- Do some tasks require footplates or use of the robotic arm?
- Are there periods of communication blackout with Mission Control that could cause problems?
- Can you always get you and your buddy back to the airlock within 30 minutes, in case of an emergency?

Although it may seem overwhelming at first, by applying logic and reasoning to the problem, you can often find a solution. As we shall see in the next section, good logic and reasoning form a vital part of the decision-making process. But before we move on to that, let's examine your powers of deduction, using a series of fun tests. Good logic and reasoning are all you need for success here (and perhaps some patience and a cup of tea!).

The scenario at the beginning of each puzzle tells you what you need to find out. Read the clues systematically, recording positive facts with a tick on the grid and negative information with a cross and making logical inferences as you go along. For example, if you were given a scenario involving coloured shapes and relative sizes and you were told, 'All the triangles, which are not the largest shapes, are blue,' you would be able to put a tick in the 'triangles/blue' box and crosses in both the 'triangles/largest' and 'blue/largest' boxes. Continue working through the clues in this fashion, then go back and cross-reference as much as you can. Fill in the answer grid at the bottom of each puzzle with any positive facts you find. All the puzzles can be solved using cross-reference, logical inference and a process of elimination. Good luck!

1. Background Information

Five would-be astronauts from varying backgrounds have submitted applications as part of the Astronaut Selection Test for the European Space Agency. Using the clues given below, can you work out each candidate's previous occupation, their hobby and their age?

Clues

1. Himmat is two years older than the software developer, who spends his or her days writing code for a pharmaceutical company.
2. Tobias is two years younger than the biology teacher.
3. Lucas, whose hobby is sailing a laser dinghy, is younger than the structural engineer.
4. The person whose hobby is rock-climbing is the youngest of the five.
5. The keen tennis player is not 32 years old.
6. The candidate who has been working as a research scientist for the past five years is older than Evie.
7. Tobias is a medical registrar at a large hospital in southern Sweden. He is two years older than the person whose hobby is gliding.
8. The 34-year-old candidate is a biology teacher.

Evie
Greta
Himmat
Lucas
Tobias

28
30
32
34
36

Astronomy
Gliding
Rock-climbing
Sailing
Tennis

Candidate	Occupation
Evie	
Greta	
Himmat	
Lucas	
Tobias	

	Biology teacher	Medical registrar	Research scientist	Software developer	Structural engineer	Astronomy	Gliding	Rock-climbing	Sailing	Tennis	28	30	32	34	36

Hobby	Age

2. Action Stations at the Station

On the International Space Station, emergency procedures are tested regularly. Using the clues, can you work out, for each module, the day a test was undertaken, the type of emergency simulated and the names of the crew members taking the principal action?

Clues

1. Tom & Jakkob were working in the Columbus module when they were faced with an emergency situation. This was neither on Friday nor on Wednesday. Friday was not the day when the crew working in the logistics module had to snap into emergency action to respond to an urgent medical situation.
2. On Tuesday, an electrical fire was the simulated emergency, but not for Benito & Saul.
3. A supposed toxic release was not the emergency situation presented to Jay & Walter or Benito & Saul.
4. Crew members Farrad & Amanda took immediate action when faced with a simulated depressurisation in either the multipurpose module or the logistics module.
5. Samantha & Bruce's responses to an emergency were tested on Monday.
6. Neither Jay & Walter nor Benito & Saul had to face an emergency in the service module. This was not the venue where a toxic release was indicated on Thursday.

Airlock

Columbus module

Logistics module

Multipurpose module

Service module

Benito & Saul

Farrad & Amanda

Jay & Walter

Samantha & Bruce

Tom & Jakkob

Depressurisation

Fire

Instrument malfunction

Medical emergency

Toxic release

Module	Day of the week
Airlock	
Columbus module	
Logistics module	
Multipurpose module	
Service module	

	Monday	Tuesday	Wednesday	Thursday	Friday	Depressurisation	Fire	Instrument malfunction	Medical emergency	Toxic release	Benito & Saul	Farrad & Amanda	Jay & Walter	Samantha & Bruce	Tom & Jakkob

Emergency	Crew members

3. Space Ventures

During extravehicular activity (EVA or spacewalks), crew members are required to complete a variety of tasks. The details below describe some of the work done on spacewalks at various times, by crew members from a few of the countries represented at the International Space Station. Using the clues, can you work out each crew member's nationality, the duration of his or her EVA and one of the tasks that he or she performed?

Clues

1. Barbara's EVA lasted 6 hours 20 minutes.

2. The longest EVA of the five was not the one undertaken by Will, who was from either the United Kingdom or the United States.

3. The task of installing a new foot restraint was part of an EVA lasting 6 hours 10 minutes, which was not undertaken by a crew member from the United States.

4. As part of his or her EVA, the crew member from Japan inspected the small robotic arm, which plays a vital part in the conducting of future experiments.

5. Carla's EVA was 2 hours 15 minutes shorter than that of the crew member who carried out maintenance work, lubricating a latching end effector. Carla is from Canada.

6. Ben carried out repair work – fixing a broken cooling loop – on his EVA.

7. The EVA lasting 5 hours 20 minutes was undertaken not by Anders, but by a crew member from Belgium.

Anders
Barbara
Ben
Carla
Will

Experiments
Get-ahead tasks
Installing equipment
Maintenance
Repair work

5 hours 15 mins
5 hours 20 mins
6 hours 10 mins
6 hours 20 mins
7 hours 30 mins

Crew member	Nationality
Anders	
Barbara	
Ben	
Carla	
Will	

	Belgium	Canada	Japan	United Kingdom	United States	5 hours 15 mins	5 hours 20 mins	6 hours 10 mins	6 hours 20 mins	7 hours 30 mins	Experiments	Get-ahead tasks	Installing equipment	Maintenance	Repair work

Duration of EVA	Type of task

4. Lunar Retrieval

There has been a crash landing on the Moon. Part of the mid-deck stowage area has detached from the spacecraft and broken up, scattering various items over a wide area. Five crew members have been despatched to retrieve as much as they can. Using the clues, can you work out, for each of the five valuable items, who found it, how many hours it took to be located and what happened to it subsequently?

Clues

1. Jens finally found the item that took longest to be found.
2. The item that was found after 24 hours of hard searching was eventually transferred to the service module.
3. John finally found the spare mini-cams, in half the time it took for the case of dehydrated food to be located. The food, its packaging intact, was used within days.
4. The spare short tethers, which were returned by neither Judi nor Jamie, were brought back before the pistol-grip tools, which were not stowed away, unused.
5. The item found by Judi was lodged behind a curiously shaped rock and took longer to be located than the item – not found by Josie – that was so badly damaged it had to be consigned to the rubbish store; but not as long as the spare oxygen tanks.

Case of food

Spare mini-cams

Spare oxygen tanks

Tethers

Tools

Scheduled for repair

Stored as rubbish

Stowed away, unused

Transferred

Used quickly

12 hours

15 hours

21 hours

24 hours

30 hours

Item	Crew member
Case of food	
Spare mini-cams	
Spare oxygen tanks	
Tethers	
Tools	

	Jamie	Jens	John	Josie	Judi	12 hours	15 hours	21 hours	24 hours	30 hours	Scheduled for repair	Stored as rubbish	Stowed away, unused	Transferred	Used quickly

No. of hours	Outcome

5. A Roving Brief

On a visit to the Moon, five trips were taken using a roving vehicle. Carrying two occupants, the vehicle was used to carry equipment and samples. Using the clues, can you identify the reason for undertaking each trip, the average speed achieved by the vehicle and the time away from the spacecraft?

Clues

1. The crew members carrying out work on Trip 3, which did not involve the collection of geological data, successfully completed their tasks and returned to the spacecraft in 3 hours. The trip was not accomplished with an average speed of 9.0 km/h, which was, however, the average speed of the trip where repair work was carried out on solar wind equipment – a trip that did not take 3 hours 15 minutes.
2. Trip 2 was the one where a deployment of solar arrays, for use in a future experiment, was effected; but it was not the trip where an average speed of 8.4 km/h was recorded.
3. The trip registering an average speed of 8.7 km/h was not the one where scientific instruments were placed in carefully selected locations.
4. An average speed of 8.9 km/h was recorded on the trip that lasted 3 hours 20 minutes.
5. A trip lasting 3 hours 50 minutes, where rock samples were collected, was not Trip 4, where the average speed was 8.4 km/h.
6. Trip 5 did not include the repairing of equipment.
7. An average speed of 8.9 km/h was not registered on Trip 1.

Trip 1	
Trip 2	
Trip 3	
Trip 4	
Trip 5	
3 hours	
3 hours 15 min	
3 hours 20 min	
3 hours 40 min	
3 hours 50 min	
8.4 km/h	
8.7 km/h	
8.9 km/h	
9.0 km/h	
9.2 km/h	

Trip	Reason
Trip 1	
Trip 2	
Trip 3	
Trip 4	
Trip 5	

	Collecting data	Collecting samples	Deploying instruments	Deploying solar arrays	Repairing equipment	8.4 km/h	8.7 km/h	8.9 km/h	9.0 km/h	9.2 km/h	3 hours	3 hours 15 mins	3 hours 20 mins	3 hours 40 mins	3 hours 50 mins

Average speed	Duration

6. Supply Issues

An unexpected delay in the sorting of various cargo shipments has occurred, as a result of a logistical mix-up on the International Space Station. Cargo from five spacecraft needs to be sorted. Using the clues, can you work out, for each item of cargo, the spacecraft it arrived on, the docking port, the date it arrived and the location where it will eventually be stored?

Clues

1. A communications equipment package arrived at the International Space Station later than the cargo docked at Node 2 (Nadir) and earlier than the date Soyuz arrived. The communications equipment was delivered by neither Progress nor HTV and was eventually placed on the external stowage platform.
2. A consignment of mass spectrometers arrived on a spacecraft that docked at Pirs. This was not on 26 July.
3. The eagerly awaited earth-sensing equipment arrived at the ISS on Soyuz.
4. Dragon brought a cargo that would be stored on the Japanese logistics module.
5. HTV docked at Node 1 on a date later than the day the spacecraft carrying the pump assembly arrived, but earlier than the arrival of the cargo destined for the functional cargo block.
6. The cargo that would eventually be stored in the multipurpose module arrived on 15 July, and Cygnus did not arrive on 18 July.
7. Crew members in the US lab eventually stored away the cargo from the spacecraft docking at the service module.

Communications equipme
Earth-sensing equipment
Food
Mass spectrometers
Pump assembly
External stowage platform
Functional cargo block
Japanese logistics module
Multipurpose module
US lab
Node 1
Node 2 (Nadir)
Node 2 (Zenith)
Pirs
Service module
7 July
10 July
15 July
18 July
26 July

Item of cargo	Spacecraft	Arrival date
Communications equipment		
Earth-sensing equipment		
Food		
Mass spectrometers		
Pump assembly		

	Cygnus	Dragon	HTV	Progress	Soyuz	7 July	10 July	15 July	18 July	26 July	Node 1	Node 2 (Nadir)	Node 2 (Zenith)	Pirs	Service module	External stowage platform	Functional cargo block	Japanese logistics module	Multipurpose module	US lab

Docking port	Storage destination

7. Spare Time in Space

Although crew members on the International Space Station spend at least two hours each day working out, to make sure muscle and bone mass are preserved, there are still opportunities for relaxation, usually at weekends. Using the clues, can you work out how five crew members spent their first hour of leisure time one day, the number of photographs they each took out of a window, to whom they communicated on the ground and which was their favourite piece of exercise equipment?

Clues

1. A TV crew arranged to record an interview with Fingal as it was his birthday. His favourite piece of exercise equipment was not the Russian treadmill.
2. Felipe, who always enjoyed the resistive-exercise device the most, did not take three photographs that day.
3. Frank took a photograph of the magnificent views twice in the first hour and told himself that was enough for the day. He did not spend his time playing his guitar.
4. Fiona's only siblings are male.
5. Reading was the first leisure activity of the day for the crew member who later spoke to her sister. This was not the crew member who preferred the VELO ergometer bike and whose first activity was listening to a Beethoven sonata and who did not communicate with her friends that day.
6. Fay, who took more photographs than Fiona did, would, if asked, nominate the T2 treadmill as her favourite piece of exercise equipment.
7. The person who started his day by playing a computer chess game took more photographs than Fay, but not as many as the crew member who spoke to his parents later that day.

Fay

Felipe

Fingal

Fiona

Frank

T2 treadmill

Cycle ergometer

Resistive-exercise dev

Russian treadmill

VELO ergometer bike

Children

Friends

Parents

Sister

TV crew

One photograph

Two photographs

Three photographs

Four photographs

Five photographs

Crew member	Type of activity	Number of photogra
Fay		
Felipe		
Fingal		
Fiona		
Frank		

Listening to music	Playing computer chess	Reading	Watching a film	Playing the guitar	One photograph	Two photographs	Three photographs	Four photographs	Five photographs	Children	Friends	Parents	Sister	TV crew	T2 treadmill	Cycle ergometer	Resistive-exercise device	Russian treadmill	VELO ergometer bike

Communication	Exercise

8. Probes' Progress

Five robotic probes owned by different countries were sent to a variety of destinations. A junior science investigator is compiling a database of the probes' accomplishments and details. Using the clues, can you help him identify each probe's destination, the mission duration, the type of samples/data collected and the country of ownership?

Clues

1. The mission lasting 7 years 3 days was undertaken by a probe owned by the Netherlands.
2. Orestes 3 flew to one of Jupiter's moons, Europa.
3. The probe owned by the Netherlands was not Macedon XI, which was not owned by Russia and did not bring back evidence of the existence of oceans.
4. Jetta 5, which was owned by Italy, did not travel to one of Neptune's moons, Proteus.
5. The duration of Jetta 5's mission was not 14 years 212 days, but Delius 2's mission did last 4 years 200 days.
6. The probe from Norway collected unusual organic compounds. This was not the probe, which was neither Jetta 5 nor Orestes 3, that completed its mission to a moon orbiting Saturn, Atlas, in a time of 3 years 42 days.
7. The probe that travelled to Phobos sent back magnificent images of its surface as it orbited Mars.
8. Spain owned the probe that travelled to Venus. This did not collect traces of metals and minerals.
9. Callinus 4 collected extremely useful solar-wind data.

Callinus 4

Delius 2

Jetta 5

Macedon XI

Orestes 3

Italy

Netherlands

Norway

Russia

Spain

Evidence of oceans

Metals and minerals

Organic compounds

Solar-wind data

Surface images

2 years 20 days

3 years 42 days

4 years 200 days

7 years 3 days

14 years 212 days

Probe	Destination	Mission duration
Callinus 4		
Delius 2		
Jetta 5		
Macedon XI		
Orestes 3		

	Atlas	Europa	Phobos	Proteus	Venus	2 years 20 days	3 years 42 days	4 years 200 days	7 years 3 days	14 years 212 days	Evidence of oceans	Metals and minerals	Organic compounds	Solar-wind data	Surface images	Italy	Netherlands	Norway	Russia	Spain

Samples/data	Country

9. Mars Manoeuvres

Sometime in the foreseeable future, on five separate occasions, astronauts may undertake missions by roving vehicle on the surface of Mars. Using the clues, can you work out, for each two-person crew, the task they will be asked to perform, how far away from the spacecraft they will be, the time their task will take and their general location?

Clues

1. Mike & Peta will travel 1 km to reach the location for their mission, which will not be to the Eridania region of Mars; neither is this the location for Brett & Christoffer's mission.
2. As they travel 4 km to collect rock samples, one crew's mission will ultimately take 2 hours 15 minutes.
3. The mission that will be based in the Thaumasia region of Mars will last for 4 hours 35 minutes and will require the crew to study their terrain for evidence of forms of life.
4. The task of one of the crews – one of those whose members have the same initial letter for their names – will be to collect experimental data. Their mission will take 3 hours 12 minutes.
5. Vinny & Juanita will be recording images on their mission to Amenthes, the duration of which will not be 2 hours 40 minutes.
6. Klaus & Kevin will travel 1 km further than the crew who will be taking readings from experimental environmental equipment, but 1 km less than the crew whose mission will be based in the Coprates region.
7. The location of one of the crews whose members have the same initial letter for their names will be the Mariner Valley. Their mission will be to examine the ground for traces of what could once have been water.

Brett & Christoffer
Klaus & Kevin
Mike & Peta
Nic & Nadiya
Vinny & Juanita
Amenthes
Coprates
Eridania
Mariner Valley
Thaumasia
2 hours 15 mins
2 hours 40 mins
3 hours 12 mins
3 hours 22 mins
4 hours 35 mins
1 km
2 km
3 km
4 km
5 km

Crew	Task	Distance
Brett & Christoffer		
Klaus & Kevin		
Mike & Peta		
Nic & Nadiya		
Vinny & Juanita		

	Collect experimental data	Collect rock samples	Consider possibility of life	Examine traces of water	Record images	1 km	2 km	3 km	4 km	5 km	2 hours 15 mins	2 hours 40 mins	3 hours 12 mins	3 hours 22 mins	4 hours 35 mins	Amenthes	Coprates	Eridania	Mariner Valley	Thaumasia

Duration	Location

10. Life on Mars

Sometime in the distant future five astronauts on Mars may continue the process of establishing a sustainable habitation. Being experienced space travellers, they each have a specialism, but the tasks they are required to perform are many and varied. Using the clues, can you work out each astronaut's area of expertise, the number of missions each has undertaken, the general task they will perform that day and the materials or an item of equipment they will help to unload?

Clues
1. The person whose task is to install a solar-energy device has flown 19 missions.
2. The unloading and stowing of a pack of spare ergometer bike seats is the responsibility of the astronaut who tackles the task of checking the growth of batches of salad vegetables.
3. Claudette, who, in the course of her working life, has become an expert on power systems, is not there when a consignment of medication is unloaded. She has completed the highest number of missions.
4. Lionel's knowledge of docking ports does not really help him when he oversees the delivery and placement of soil-collection materials. He has completed fewer missions than Jim.
5. Noah, who does not unload any medication, has completed four fewer missions than the crew member who helps to unload a shipment of scientific instruments, but more missions than the astronaut whose task of the day is concerned with vital oxygen generation.
6. Serena, whose area of expertise is not in the field of environmental control, makes a good contribution to the complicated task of water recovery.
7. The main task of the day for the astronaut who is known to be extremely interested in airlock design is the removal and repair of a nut on the rear wheel of a roving vehicle.

Claudette

Jim

Lionel

Noah

Serena

Exercise equipment

Medication

Scientific instruments

Soil-collection materia

Water-extraction mate

Energy-device installa

Farming

Oxygen generation

Vehicle repair

Water recovery

11 missions

15 missions

17 missions

19 missions

21 missions

Astronaut	Area of expertise	Number of missions
Claudette		
Jim		
Lionel		
Noah		
Serena		

Airlocks	Docking ports	Environmental control	Life-support systems	Power systems	11 missions	15 missions	17 missions	19 missions	21 missions	Energy-device installation	Farming	Oxygen generation	Vehicle repair	Water recovery	Exercise equipment	Medication	Scientific instruments	Soil-collection materials	Water-extraction materials

Task	Equipment/materials

Answers

Space logic puzzle 1: Background Information

Answers

Evie	Software developer	Rock-climbing	28
Greta	Structural engineer	Tennis	36
Himmat	Research scientist	Gliding	30
Lucas	Biology teacher	Sailing	34
Tobias	Medical registrar	Astronomy	32

Explanation

The candidate who is 28 years old and whose hobby is rock-climbing (clue 4) cannot be the structural engineer (clue 3) or the research scientist (clue 6) or the medical registrar, Tobias (clue 7). As the biology teacher is 34 (clue 8), the 28-year-old rock climber must be the software developer, which means Himmat is 30 (clue 1) and Lucas (clue 3) is at least 32, therefore the structural engineer must be 36 (as the biology teacher is 34 (clue 8) and Himmat is the research scientist). Evie must therefore be the software developer and Tobias must be 32 years old. This means gliding is Himmat's hobby (clue 7).

As Lucas is not a structural engineer (clue 3), this must be Greta's occupation and she is therefore 36 years old, leaving Lucas as the biology teacher who sails. Tobias's hobby is not tennis (clue 5), so it must be astronomy, leaving Greta as the tennis player.

Space logic puzzle 2: Action Stations at the Station

Answers

Airlock	Tuesday	Fire	Jay & Walter
Columbus	Thursday	Toxic release	Tom & Jakkob
Logistics	Wednesday	Medical	Benito & Saul
Multipurpose	Friday	Depressurisation	Farrad & Amanda
Service	Monday	Instrument malfunction	Samantha & Bruce

Explanation

Tom & Jakkob were the crew members responding to an emergency in Columbus (clue 1). Farrad & Amanda were faced with an emergency in either the multipurpose or the logistics modules (clue 4). Neither Jay & Walter nor Benito & Saul were faced with an emergency in the service module (clue 6), therefore Samantha & Bruce must have been the crew facing an emergency in the service module, which must have been on Monday (clue 5).

Samantha & Bruce cannot therefore have been presented with a medical situation, as this took place in the logistics module (clue 1); a fire, as this was on Tuesday (clue 2); a depressurisation situation, as this was the emergency presented to Farrad & Amanda (clue 4); or a toxic release, as this happened on Thursday (clue 6). This means their emergency must have been an instrument malfunction.

As the toxic-release emergency was not presented to either Jay & Walter or Benito & Saul (clue 3), and Farrad & Amanda were faced with a depressurisation situation (clue 4), a toxic release must have been the emergency presented to Tom & Jakkob, in the Columbus module (clue 1), on Thursday (clue 6).

The medical-emergency simulation took place in the logistics module (clue 1), which means that Farrad & Amanda's simulated depressurisation took place in the multipurpose module (clue 4).

Benito & Saul were not faced with a simulated fire (clue 2), which means their emergency must have been a medical one, in the logistics module (clue 1), leaving Jay & Walter facing a fire emergency in the airlock on Tuesday. As the medical emergency did not happen on Friday (clue 1), it must have taken place on Wednesday, leaving Friday as the day of the depressurisation simulation.

Space logic puzzle 3: Space Ventures

Answers

Anders	United States	7 hours 30 mins	Maintenance
Barbara	Japan	6 hours 20 mins	Experiments
Ben	Belgium	5 hours 20 mins	Repair work
Carla	Canada	5 hours 15 mins	Get-ahead tasks
Will	United Kingdom	6 hours 10 mins	Installing equipment

Explanation

Maintenance was not the task carried out by Carla (clue 5) or Ben (clue 6), and as the task took 7 hours 30 minutes (clue 5), it was neither Will (clue 2) nor Barbara (clue 1), leaving Anders as the one with the maintenance task.

Anders's task therefore took 7 hours 30 minutes (clue 5), Barbara's task took 6 hours 20 minutes (clue 1), Carla's took 5 hours 15 minutes (clue 5), and as the task taking 5 hours 20 minutes was completed by a crew member from Belgium (clue 7), it could not have been Will (clue 2), therefore the task taking 5 hours 20 minutes was completed by Ben, leaving Will's task taking 6 hours 10 minutes.

Will's task was therefore the installation task (clue 3), and as this task was not completed by a crew member from the United States (clue 3), and Will was from either the United States or the United Kingdom (clue 2), he must be from the United Kingdom. Carla is from Canada (clue 5), and so Ben, with a task taking 5 hours 20 minutes, must be from Belgium (clue 7).

As Anders's task is maintenance, Ben's is repair work and Will's is installing equipment, the work on experiments must be the task of either Carla or Barbara. This task was carried out by a crew member from Japan (clue 4), so it must have been Barbara's, leaving Carla as the one completing the get-ahead work and, by process of elimination, making Anders the United States crew member.

Space logic puzzle 4: Lunar Retrieval

Answers

Case of food	Jens	30 hours	Used quickly
Spare mini-cams	John	15 hours	Stored as rubbish
Spare oxygen tanks	Jamie	24 hours	Transferred
Tethers	Josie	12 hours	Stowed, unused
Tools	Judi	21 hours	Scheduled for repair

Explanation
The spare mini-cams were found by John in half the time it took for the food to be found (clue 3). This means it must have taken either 30 hours or 24 hours for the food to be found. As we know the food was used quickly (clue 3) and the item that was found after 24 hours was transferred to the Service module (clue 2), the food must have

been found after 30 hours, which means John found the mini-cams after 15 hours and Jens found the food (clue 1).

As the tethers were not found by Judi or Jamie (clue 4), it must have been Josie who found them. Judi did not find the spare oxygen tanks (clue 5), therefore she must have found the tools, leaving Jamie locating the spare oxygen tanks.

Neither the oxygen tanks (clue 5) nor the tools (clue 4) were found in the shortest time, so Josie must have found the tethers after 12 hours. This means the oxygen tanks must have been found after 24 hours and transferred to the Service module (clues 5 and 2), and Judi must have found the tools after 21 hours.

The tools were not consigned to the rubbish store (clue 5) or stowed away, unused (clue 4), so they must have been scheduled for repair. The tethers found by Josie were not stowed as rubbish (clue 5), so they must have been stored away, unused, which means the spare mini-cams were the items to be stowed in the rubbish store.

Space logic puzzle 5: A Roving Brief

Answers

Trip 1	Repairing equipment	9.0 km/h	3 hours 40 mins
Trip 2	Deploying solar arrays	8.9 km/h	3 hours 20 mins
Trip 3	Deploying instruments	9.2 km/h	3 hours
Trip 4	Collecting data	8.4 km/h	3 hours 15 mins
Trip 5	Collecting samples	8.7 km/h	3 hours 50 mins

Explanation

The trip with an average speed of 8.9 km/h had a duration of 3 hours 20 minutes (clue 4) and the trip for the collection of rock samples took 3 hours 50 minutes (clue 5), so the trip for repairing equipment, which was where an average speed of 9.0 km/h was registered (clue 1) and was of neither 3 hours nor 3 hours 15 minutes' duration (clue 1), must have lasted for 3 hours 40 minutes.

Trip 2 was the one where solar arrays were deployed (clue 2), and the average speed for Trip 4 was 8.4 km/h (clue 5), so the trip for the repairing of equipment, which was neither Trip 3 (clue 1) nor Trip 5 (clue 6), must have been Trip 1.

Trip 3 took 3 hours (clue 1), so Trip 4, at 8.4 km/h and which was not of 3 hours 50 minutes' duration (clue 5), must have taken 3 hours 15 minutes.

By process of elimination, Trip 2 must have taken 3 hours 20 minutes to deploy solar arrays, with an average journey speed of 8.9 km/h, which means Trip 5 took 3 hours 50 minutes and its purpose was to collect rock samples.

Trip 3 was not for collecting data (clue 1), so it must have been for deploying scientific instruments, which leaves collecting data as the purpose of Trip 4, at a duration of 3 hours 15 minutes.

8.7 km/h was not the average speed of the trip for deploying scientific instruments (clue 3), so Trip 3 must have had an average speed of 9.2 km/h, which means Trip 5 must have had an average speed of 8.7 km/h.

Space logic puzzle 6: Supply Issues

Answers

Comms. equip.	Cygnus	10 July	Node 2 (Zenith
Earth-sensing equip.	Soyuz	26 July	Service module
Food	HTV	15 July	Node 1
Mass spectrometers	Progress	18 July	Pirs
Pump assembly	Dragon	7 July	Node 2 (Nadir)

Explanation

The earth-sensing equipment was delivered by Soyuz (clue 3), and neither Progress nor HTV brought the communications equipment, which means it was delivered by either Cygnus or Dragon, but the communications equipment was stowed on the external stowage platform (clue 1) and Dragon's cargo was stored in the Japanese logistics module (clue 4), so the communications equipment must have been delivered by Cygnus.

Cygnus cannot have docked at Node 2 (Nadir) (clue 1) or the service module (clue 7) or Node 1 (clue 5) or Pirs (clue 2), so it must have docked at Node 2 (Zenith). The mass spectrometers were delivered by a spacecraft docking at Pirs (clue 2). This cannot have been Cygnus or Soyuz or HTV (clue 5), so it must have been either Progress or Dragon. As HTV's cargo was not stowed on the external stowage platform, the US lab, the Japanese logistics module or the functional cargo block, it must have been stowed at the multipurpose module. We now know that its cargo cannot have been the pump assembly (clue 5), the communications equipment, the earth-sensing equipment or the mass spectrometers, which leaves food as the cargo that HTV delivered to Node 1, which was stowed in the multipurpose module on 15 July.

External stowage platform
US lab
Multipurpose module
Functional cargo block
Japanese logistics module

Cygnus docked at Node 2 (Zenith), HTV at Node 1, and Soyuz did not dock at Node 2 (Nadir) (clue 1), which means it either docked at Pirs or the service module. The cargo delivered to Pirs was mass spectrometers, which means Soyuz docked at the service module with its cargo of earth-sensing equipment, which was stowed at the US lab (clue 7).

Neither the food on HTV nor the pump assembly would be stored at the functional cargo block (clue 5), leaving the mass spectrometers as the cargo to be stored there, and the Japanese logistics module as the destination for the pump assembly, which would be delivered by Dragon, leaving Progress as the spacecraft delivering the mass spectrometers and docking at Pirs, and Dragon docking at Node 2 (Nadir).

We can now use two clues to plot the sequence of arrivals:
Clue 1: Dragon . . . Cygnus . . . Soyuz
Clue 5: Dragon . . . HTV . . . Progress

We know HTV arrived on 15 July (clue 6) and Cygnus did not arrive on 18 July (clue 6). As Progress was not the last spacecraft to arrive (clue 2) and Dragon is the earliest arrival in both sequences, we can infer that Dragon arrived on 7 July, Cygnus on 10 July, HTV on 15 July, Progress on 18 July and Soyuz on 26 July.

Space logic puzzle 7: Spare Time in Space

Answers

Fay	Reading	3 photos	Sister
Felipe	Playing the guitar	5 photos	Parents
Fingal	Computer chess	4 photos	TV crew
Fiona	Listening to music	1 photo	Children
Frank	Watching a film	2 photos	Friends

Explanation

Fiona did not speak with a sister (clue 4). The crew member who did was female (clue 5) and therefore must have been Fay. Fingal spoke with a TV crew (clue 1). The person who spoke with parents was male (clue 7), which means that Fiona's conversation, as the other female and the one who didn't speak with friends (clue 5), must have been with her children.

Frank took two photos (clue 3), and Fay took more than Fiona (clue 6). We know there were two crew members who took more photos than Fay (clue 7), so Fay must have taken three photos, leaving Fiona as the one who took one; the crew member playing computer chess as the one who took four photos; and the person who later spoke to parents, five photos (clue 7).

We now know that Fay's activity was reading and Fiona's was listening to music (clue 5). Frank did not play the guitar (clue 3), neither did he play computer chess, as the person who did so took four photos, leaving Frank as the crew member who watched a film.

2 treadmill
esistive-exercise device
ycle ergometer
ELO ergometer bike
ussian treadmill

Fiona's favourite exercise equipment was the VELO ergometer bike (clue 5). Fay preferred the T2 treadmill (clue 6) and Felipe the resistive-exercise device (clue 2). Frank could not have been the crew member who spoke to parents (clue 7), which means he must have had a conversation with friends, leaving Felipe as the person speaking to his parents and therefore taking five photos; Fingal as the player of computer chess who took four photos; and Felipe as the player of the guitar. As his favourite exercise equipment was not the Russian treadmill (clue 1), Fingal must have preferred the cycle ergometer, leaving the Russian treadmill as Frank's favourite.

Space logic puzzle 8: Probes' Progress

Answers

Callinus 4	Atlas	3 years 42 days	Solar-wind data
Delius 2	Venus	4 years 200 days	Evidence of oceans
Jetta 5	Phobos	2 years 20 days	Surface images
Macedon XI	Proteus	14 years 212 days	Organic compounds
Orestes 3	Europa	7 years 3 days	Metals and minerals

Explanation

Jetta 5 is owned by Italy (clue 4), the mission lasting 7 years 3 days is a Netherlands initiative (clue 1) and the mission to Venus is controlled by Spain (clue 8), so the Atlas mission, which lasted 3 years 42 days and was not owned by Norway (clue 6), must have been owned by Russia. Orestes 3 travelled to Europa (clue 2), so Jetta 5, which because it is owned by Italy cannot have been sent to Venus or Atlas and was not travelling to Proteus (clue 4), must have gone to Phobos and was therefore sent to collect surface images (clue 7).

Jetta 5's mission did not last for 14 years 212 days and Delius 2's mission lasted 4 years 200 days (clue 5), so the duration of Jetta 5's mission must have been 2 years 20 days. Russia's probe that travelled to Atlas is not Macedon XI (clue 3), so its probe must have been Callinus 4, on a mission that lasted 3 years 42 days, and it must have collected the solar-wind data (clue 9).

Russia

Spain

Italy

Norway

Netherlands

Norway's probe collected organic compounds (clue 6), and as the Spain probe travelled to Venus and did not collect metals and minerals (clue 8), the Spanish probe must have collected evidence of oceans. This means the Macedon XI probe must have travelled to Proteus, and the metals and minerals must have been collected by the probe owned by the Netherlands on a mission that lasted 7 years 3 days. This cannot have been the Macedon XI probe (clue 3), so it must have been Orestes 3, on a mission to Europa (clue 2), which means the Macedon XI probe, owned by Norway, collected organic compounds on a mission lasting 14 years 212 days.

Space logic puzzle 9: Mars Manoeuvres

Answers

Brett & Christoffer	Rock samples	4 km	2 hours 15 mins
Klaus & Kevin	Traces of water	3 km	2 hours 40 mins
Mike & Peta	Possibility of life	1 km	4 hours 35 mins
Nic & Nadiya	Experimental data	2 km	3 hours 12 mins
Vinny & Juanita	Record images	5 km	3 hours 22 mins

Explanation

Either Klaus & Kevin or Nic & Nadiya will be collecting experimental data on a mission lasting 3 hours 12 minutes (clue 4). This cannot be Klaus & Kevin (clue 6), so it must be Nic & Nadiya, which means Klaus & Kevin's mission must be the one to Mariner Valley, where they will search for traces of water (clue 7). Vinny & Juanita will be recording images (clue 5). As the crew on the mission to collect rock samples will travel 4 km (clue 2) and Mike & Peta will travel 1 km on their mission, the crew collecting rock samples – on a mission lasting 2 hours 15 minutes (clue 2) – must be Brett & Christoffer, leaving Mike & Peta as the crew travelling to Thaumasia to search for possible life signs on a mission lasting 4 hours 35 minutes (clue 3).

We know Klaus & Kevin's mission is to the Mariner Valley, Mike & Peta's to Thaumasia and Vinny & Juanita's is to Amenthes (clue 5). As Brett & Christoffer will not travel to Eridania (clue 1), their mission must be to Coprates, leaving Nic & Nadiya as the crew travelling to Eridania. As Mike & Peta will travel 1 km and Brett & Christoffer 4 km to Coprates (clue 2), this means Klaus & Kevin must travel 3 km to Mariner Valley and Nic & Nadiya 2 km to Eridania (clue 6), leaving Vinny & Juanita to travel 5 km on their mission.

oprates
ariner Valley
haumasia
ridania
menthes

We are told Vinny and Juanita's mission will not take 2 hours
40 minutes, which means that must be the duration of Klaus &
Kevin's mission, leaving the duration of Vinny & Juanita's mission
as 3 hours 22 minutes.

Space logic puzzle 10: Life on Mars

Answers

Claudette	Power systems	21	Farming
Jim	Environment control	19	Energy device
Lionel	Docking ports	11	Oxygen generatio
Noah	Airlocks	15	Vehicle repair
Serena	Life-support systems	17	Water recovery

Explanation

Claudette is an expert on power systems (clue 3), Lionel's area of expertise is docking ports and he unloads the soil-collection materials (clue 4), and the task of the astronaut who is interested in airlock design is to repair a vehicle (clue 7). As Serena's task is concerned with water recovery and she is not an expert in environmental control (clue 6), her area of expertise must be life-support systems.

Claudette has completed 21 missions (clue 3), and the person given the energy-device installation task has completed 19 missions (clue 1). Claudette cannot have taken on the vehicle-repair task (clue 7) and Serena's task is water recovery (clue 6), meaning that Claudette's task must be either oxygen generation or farming. The oxygen-generation task must have been accomplished by someone with fewer than 21 missions (clue 5), leaving Claudette with the farming task and the responsibility for unloading the exercise equipment (clue 2).

We are told Noah has completed four fewer missions than the person unloading the scientific instruments (clue 5). This cannot be Claudette. Therefore the number of missions completed by the person unloading scientific instruments must be either 19 or 15, leaving Noah as having had 11 or 15 missions, but as Noah has

Exercise equipment
Scientific instruments
Soil-collection materials
Water-extraction materials
Medication

completed more missions than the person responsible for oxygen generation (clue 5), his number of missions cannot be 11. This means the sequence must be oxygen generation 11 missions, Noah 15 missions and the unloader of scientific instruments 19 missions. Noah's task, as he has 15 missions, is therefore vehicle repair (given that Claudette's task is farming and Serena's is water recovery, and the energy-device installation task is undertaken by the person completing 19 missions) and his field of expertise must be airlocks, leaving Jim as the person with knowledge of environmental control.

As Noah, with 15 missions, does not unload the medication (clue 5) or scientific instruments, the soil-collection materials are unloaded by Lionel (clue 4) and exercise equipment by Claudette, Noah must unload the water-extraction materials. Serena's task is water recovery, which means – as Lionel has fewer missions than Jim (clue 4) – that Jim's task must be installing the energy device, he must have completed 19 missions and must unload the scientific instruments, leaving Lionel with 11 missions and the task of oxygen generation, and Serena with 17 missions and the responsibility for the unloading of the medication.

DECISION-MAKING

During the final interview of astronaut selection I was asked what one quality an astronaut should possess, above all others. After a brief moment I answered: 'good judgement'. Judgement is essentially the ability to make considered decisions or come to sensible conclusions. This is easy to say, much harder to practise. Some people are inherently good decision-makers and may not even consciously be aware of the processes that are leading them to a sensible conclusion.

Good judgement, or decision-making, is the universal key to success. As such, it is a much-studied subject – and the good news is that, for those people who struggle to find solutions to complex problems, it is a skill that can be learnt. There are many different techniques and examples showing how to approach a problem and arrive at the best solution. These processes vary slightly; a model used by the military, for example, may not be identical to one used in aviation or the medical profession. However, the end goal is the same: sound judgement.

During training, ESA astronauts are taught a six-stage process for decision-making and problem-solving. The six steps (in alphabetical order) are: **C**heck, **D**ecide, **E**xecute, **F**acts, **O**ptions, **R**isks. This process is known by a six-letter acronym made from the letters in bold type. Using good logic, can you work out the appropriate order of the steps and identify which of the following is the correct acronym? The answer is at the top of the next page.

a) FORCED
b) DRECOF
c) FORDEC
d) CORFED

The aswer is FORDEC – one method of applying structure and logic to a complex problem. Here's how it works.

- **F**acts: First you need to collect all the relevant information about a problem. You should also be careful to assess any details that are missing and search them out. Here's an example. According to instrument readings, an antenna has failed to retract. First the crew would confirm that status reading with the ground and by other means, and then collect information on the impact of the problem on other planned operations.

- **O**ptions: Once the facts have been gathered, it is time to list possible ways to resolve the problem. You should take into account past experience when considering these options. In the case of the antenna, the crew and the ground might discuss whether the problem can be solved during an EVA, repeating the retraction procedure or repairing a software error, and so on. If an EVA is the only solution, can the task be done during a planned EVA or is a special one necessary? Generating options can be an art in itself. Too few options and you may need to think about the problem from a different angle or get fresh ideas from another person. Too many options and the decision-making process is going to become laborious and time-consuming. Selecting about three realistic options is usually a good number.

- **R**isks: Now it is time to evaluate the risks and benefits of the different options you have listed. You should identify the pros and cons of an action and consider the trade-off between them, to see if it is worth the risk. You should do so using past experience, where possible. You'll have to consider the resources you have available (including time).

With the antenna problem, the crew and ground would evaluate the pros and cons and estimate the likely success of manipulating the antenna into place for each option generated.

- **D**ecide: Based on assessing the pros and cons of each option, it's now time to decide what level of risk you are prepared to take, and then select a course of action. In the antenna example, it might be decided that the astronauts will try to close the antenna on the next scheduled spacewalk. It is important that once the decision is made, it is followed through.

- **E**xecute: Put the decision into action. The astronauts would spend time preparing the tools, evaluate their routes to the antenna and make preparations to amend their EVA schedule to include the extra work.

- **C**heck: Once a decision is executed, you should check the outcome and compare it with your initial expectation and risk/benefit estimate. If the executed action didn't solve the problem, you should reapply the FORDEC process in full, without bypassing any of the steps. For example, if the antenna still didn't retract, then the astronauts would restart the problem-solving process with the ground by reassessing the facts, perhaps collecting more visual information to help determine if the antenna is likely to interfere with future operations and what could be holding the antenna in place.

Under many circumstances in space, the FORDEC model can be applied to solve complex problems. In the example given above concerning the antenna, this process may involve many people in Mission Control with different specialities – engineering, guidance,

NEUTRAL BUOYANCY TRAINING

The principle used to simulate weightlessness in a huge tank of water is called 'neutral buoyancy'. For an astronaut to be neutrally buoyant in water, the natural tendency to float or sink is counteracted by weights or flotation devices. Although it is not exactly the same as being weightless in space, astronauts can practise in water how to perform an extravehicular activity (EVA, or 'spacewalk').

At ESA's Neutral Buoyancy Facility (NBF) in Cologne, astronauts learn basic EVA concepts and skills, such as tethering to the International Space Station, the use of special tools, communicating with an EVA crewmate and with the control room, as well as how to keep full situational awareness in a complex and challenging environment.

The NBF pool has an accompanying mock-up of the European Columbus ISS module that can be lowered into the water, allowing astronauts to practise with a spacecraft on Earth. The pool is 10 metres at its deepest, meaning there is plenty of room for the module, the astronauts and the instructors.

Real spacewalks typically last between six and eight hours. You need to have stamina and the ability to keep your concentration levels up.

Once ESA astronauts complete their basic training course in Cologne, they travel to NASA's Neutral Buoyancy Laboratory (NBL) near the Johnson Space Center in Houston, Texas. At 62 metres long, 31 metres wide and over 12 metres deep it is the largest indoor pool in the US. With a total capacity of 23.5 million litres of water, it holds an almost full-scale replica of the ISS.

At the NBL, astronauts wear full spacesuit replicas. Each suit weighs around 135 kg, so when wearing them outside of the pool we have to be attached to a metal frame and lowered into the water using a crane. Two 26-metre umbilical cables supply us with air to breathe while we work underwater.

NEEMO

Aquarius is a submerged habitat deep on the sea floor next to coral reefs, nearly 6 km off the coast of Florida. Ordinarily it is home to marine biologists, but occasionally a crew of astronauts take over as part of NASA's Extreme Environment Mission Operations (NEEMO) expedition.

Aquarius's isolation makes it a very suitable environment to prepare for future spaceflight missions. If there's an emergency, it takes 17 hours to safely decompress and return to the surface – longer than it takes to return from the International Space Station.

During our 12-day training trip to Aquarius, one of our tasks was to evaluate a crewed asteroid mission. Asteroids are much smaller than planets, meaning they have a tiny gravitational pull. On an asteroid you might weigh less than a 1kg bag of sugar does on Earth. One wrong move and you could float away into space. So we looked at ways to anchor ourselves to, navigate around and conduct scientific experiments on an asteroid.

We used a jet-pack and two deep water submersible vehicles that had foot plates attached, similar to when we have our boots strapped to a plate while performing spacewalks outside the ISS.

CENTRIFUGE TRAINING

The closest astronauts get to experiencing the feeling of launch during their training happens at Star City – the Russian space agency's training centre outside of Moscow. It is home to the world's largest centrifuge – an 18-metre arm that spins around inside a circular room. At the end of the arm is a cabin designed to mimic the inside of the Soyuz capsule, complete with display panels, knobs and dials.

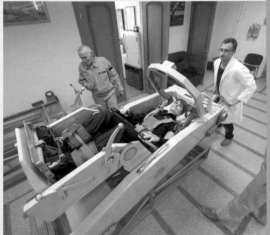

As the arm accelerates around, it creates high g-forces that make you feel much heavier than normal. Scientists talk of g-forces in terms of multiples of the force you feel due to Earth's normal downwards gravitational pull. This is said to be 1g. The higher the g-force, the more you feel pressed down into your seat. The instructors start you off gently, only taking you up to around 4g on your first run.

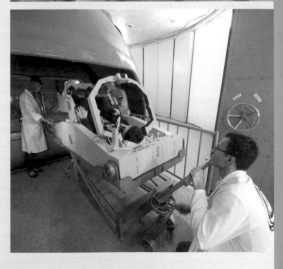

Before your first run on a centrifuge you are taught some of the basic techniques for dealing with increased g-forces. Breathing is key. You can't easily breathe through your chest because your ribcage is working very hard to withstand the high g-force. Instead, you have to 'lock' your chest in place as if you are doing a bench press, and breathe through your stomach. It takes a little bit of practice.

Once you've become accustomed to the centrifuge, the instructors put you through a Soyuz launch profile. This sees you in the centrifuge for nearly nine minutes – which feels like a long time to be withstanding high g-forces! You go through all the stages of launch, including when the rocket separates and the g-force drops, only then to increase again as you accelerate towards the ISS.

Contrary to popular belief, the centrifuge doesn't make you feel dizzy or sick. Although it does spin around, it is not like a fairground ride that may come to mind. The capsule is on a hinge, meaning you always feel the g-force through your chest exactly like you would on a Soyuz launch or re-entry. You don't even notice that you're spinning around.

ZERO-G TRAINING

To prepare for life in microgravity, astronauts spend time in the colloquially named 'Vomit Comet' – a converted Airbus jet airliner that executes a series of bell-shaped (parabolic) trajectories in the sky. The plane has space on board for 40 passengers, and for their protection there is padding on the walls, the ceiling and the floor.

After the plane takes off from a normal runway it climbs to 6,000 metres. From there, the pilot pulls up to a climbing angle of approximately 45 degrees, taking the aircraft to an altitude of around 8,500 metres and pulling around 2g in the process. The pilot starts the zero-g phase by gradually pushing forward on the control column until the aircraft is at a nose-down angle of around 45 degrees. During this zero-g period, which lasts around 25 seconds, the passengers are weightless and in freefall towards the ground. Finally, the pilot will pull out of the descent until the aircraft is once again straight and level at 6,000 metres.

During zero-g training, astronauts learn how to move their bodies in weightlessness. You have to use your arms to turn your whole body, for example. We practise hooking our feet under handrails in order to stabilise ourselves so that we can work. Learning how to drink in weightlessness is great fun, seeing the water form tiny floating droplets in front of you. That stood me in good stead for life in orbit, and I could demonstrate the unusual behaviour of water to school-children back on the ground.

A whole parabolic flight can last three and a half hours in total and astronauts typically get around 30 short bursts of weightlessness during that time. But as preparation for life on the ISS, every second is worth it.

navigation and control, EVA, and so on – and it may take several hours to reach a conclusion. However, it is also an equally valid model for making rapid decisions within a small crew.

Where time is of the essence, all the skills covered during astronaut training will come to the fore. Good situational awareness will rapidly identify the facts; logic and reasoning will help identify the options and risks; and teamwork and communication will be vital in executing the course of action.

Next time you have a difficult decision to make, try applying the FORDEC model and see if helps lead to a sensible conclusion.

PERSONALITY

What kind of personality do you have? A lot of sound self-care is about knowing yourself inside out: what your strengths and weaknesses are and how you deal best with different situations.

Here's something to try, from the psychological part of the astronaut selection process. How would you answer the following questions about yourself? You may write short remarks (in this book or on a separate piece of paper or a computer); complete sentences are not necessary.

1. What have been the major aspects in your personal development?

_____

_____

_____

_____

_____

2. How do you currently think about your secondary and tertiary education, about your classmates, teachers, colleagues and superiors?

_____

_____

_____

_____

_____

3. What roles have you played in school, university, job, clubs or other groups?

_____

_____

_____

_____

_____

4. What have been special events, experiences, acknowledgements, successes, failures, disappointments and critical situations in your life?

_____

_____

_____

_____

_____

5. What are your hobbies and personal interests?

_____

_____

6. What accidents, serious illnesses or injuries have you had?

_____

_____

_____

_____

_____

7. Indicate personal character traits that have brought you:
   a) advantages, b) disadvantages.

_____

_____

_____

_____

_____

8. What led you to apply to become an astronaut?

_____

_____

_____

_____

The way you answer these questions helps selectors and astronaut trainers assess your personality type. In one famous study, Isabel Briggs Myers and Katharine Cook Briggs built on the work of psychologist Carl Jung to suggest that we can all be split into 16 different personality types, through the so-called Myers–Briggs Test. Although astronaut selectors don't explicitly use this test, they are employing psychological tasks to assess similar qualities.

TEST 1

In the Myers–Briggs Test, a person falls into one of two possible categories for each of four key personality traits. They are:

- **I**ntroversion vs **E**xtroversion: Contrary to popular belief, this isn't about being shy or loud/exuberant. Instead it is about whether you derive your energy from being on your own or from being part of a group. Are you more stimulated by your environment or by the thoughts inside your own head?

- **S**ensing vs i**N**tuition: This one is about how you perceive things. Sensors take in the majority of their information from the five senses of smell, touch, taste, sight and hearing. By contrast, those in the intuition camp often use their 'sixth sense' or gut instinct.

- **T**hinking vs **F**eeling: How do you make decisions? Thinkers are objective and use logic to form conclusions. Feelers rely far more on subjective factors when arriving at a decision.

- **J**udging vs **P**erceiving: To establish which of these you favour, think about how you organise your life. Judgers are decisive, planned and orderly, whereas perceivers are often more flexible, adaptable and spontaneous.

Try and decide for yourself which category applies most to you in each of the four stages. You then put the four letters in bold together to form your personality type. So you could be an ISTJ, for example. Or an ENFP.

When it comes to being an astronaut, there is no one correct answer, with regard to personality type. In fact, as space agencies start to consider long-duration missions to the Moon and Mars, a more diverse crew is likely to be most successful. However, extremes of personality can be problematic, and it's good to know your own strengths and weaknesses when working with others. HBP training helps you develop these skills further.

What do you think would be some of the strengths and weaknesses of the following personality types?

a) ENTJ
b) ISFP
c) ESTP

And, perhaps more importantly, what personality types would you most like to fly in space with, as crewmates? There is no correct answer to this question – it will vary for each person, depending on their psychological make-up.

TEST DEBRIEF

As a trainee astronaut, learning Russian and HBP training form part of 18 months of basic training. During that time you also receive an overview of the major space-faring nations, of their space agencies (with special emphasis on the European Space Agency) and of the major human and robotic space programmes. You also learn about space law and intergovernmental agreements governing worldwide cooperation in space.

Next you start to get into the technical details that are crucial to launching, living and working in space. You look at spaceflight engineering, electrical engineering, aerodynamics, propulsion, orbit mechanics, materials and structures. Be prepared to tackle scientific disciplines such as research under weightlessness, Earth observation and astronomy.

Basic training also provides a detailed overview of the International Space Station and its systems. You learn about its structure and design, guidance navigation and control, thermal control, electrical power generation and distribution, command and tracking, life-support systems, robotic systems, systems for spacewalks and payload systems. Ground systems such as test sites, launch sites, training and control centres are also reviewed.

The final part of basic training covers so-called 'special skills', such as robotics and survival. Astronauts also develop SCUBA skills and learn the basic aspects of rendezvous and docking with the ISS. To read in more detail about some of these skills, turn now to the photo inset sections of this book. Otherwise, get ready for Part Four – and the future of what an astronaut's journey to space may look like: Mars training.

PART FOUR
MISSION TO MARS

MARS TRAINING

'Houston, we have a problem solver!'

Congratulations, you're now well on your way to becoming a fully fledged astronaut. The tests you have already completed have equipped you with many of the key skills and attributes that you will put to good use on a mission to the International Space Station.

However, humans possess an insuppressible desire to explore. We're always looking to reach the next horizon, to embark on the next exciting adventure. The time is fast approaching when humans will leave Low Earth Orbit again and venture out into the wider solar system. We've already been to the Moon, and a return is on the cards to establish a more permanent habitation base and as a stepping-stone to more distant destinations. But Mars really stands out now as the ultimate challenge. No other planet in our solar system presents an opportunity for humans to live, work and explore, using our current levels of technology. Mercury and Venus are too scorchingly hot; the four outer planets don't have solid surfaces.

Today Mars *is* inhabited – by a fleet of robotic landers and rovers that we have sent there to explore in our stead. They have revealed a fascinating world, with tantalising clues of an incredible history. Mars is now a dry, dusty expanse, but there is growing evidence that it had oceans of liquid water in the past. Maybe it started off similar to Earth, before the evolution of the two planets diverged significantly. It might just be possible that life started on Mars at around the same time it did here, and that life is still clinging on, somewhere on the Red Planet.

Sending humans to Mars will help us determine whether we are alone in the solar system. It will kick-start a chain of events that might one day see a permanent human presence there. If we could establish an outpost on the planet, it might offer an insurance policy for humanity, in case a disaster threatens life here on Earth. And we know that pushing the envelope of exploration will have added benefits. A trip to Mars will force us to invent creative solutions to new problems that can be adapted for the benefit of everyone – whether that means breakthroughs in medicine, technology and sustainability or the creation of a new field that we haven't even conceived of yet.

In all likelihood, the first person to walk on Mars is alive today – he or she may well currently be at school. It could be you. So what do you need in your locker, to be that person?

In order to become the first astronaut on the Red Planet you will need to possess a unique set of strengths and skills – even beyond the stringent requirements you have already encountered in this book. The brave pioneers of Mars will encounter physical and psychological stresses that are in some cases hitherto unknown to any human. But don't be afraid. Numerous experts at ESA and other space agencies around the world are already planning for such a mission. Long-duration training experiments are being conducted to see what happens to humans when we spend great periods of time in extreme environments; when we are kept apart from our families and loved ones, and are isolated with only a few fellow astronauts, untethered from civilisation.

A mission to Mars will be a challenge of a completely new magnitude, and the astronaut selection and training process explored in this book may change significantly in the future, as we learn more about what it takes to survive and thrive in these conditions of deep space.

This final part of the book includes questions, activities, puzzles and interviews based on the latest research into a crewed mission to Mars. You are already among the elite. Now get ready to join the Martian elite. Good luck, Astronaut.

COMMENCING FINAL TEST PROCEDURE

TEST 1: Pattern recognition

Below you will see a series of shapes and patterns. If the shapes or patterns are arranged in a line, you will have to work out the next object in the sequence, choosing from several options. If the shapes or patterns are arranged in a grid, you will have to choose which option should fill the missing square. You will have to be quick, though, as you're only allowed ten seconds for each puzzle.

Here is an example sequence:

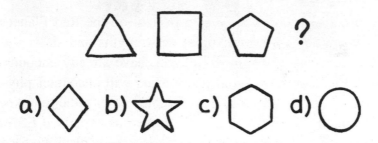

The answer would be c) because each consecutive shape has one extra side.

Now try to complete the following pattern-recognition tests:

1.

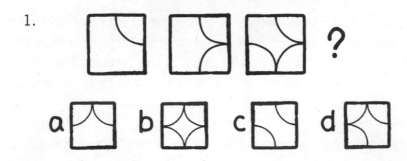

2.

a   b   c   d

3.

a   b   c   d

4.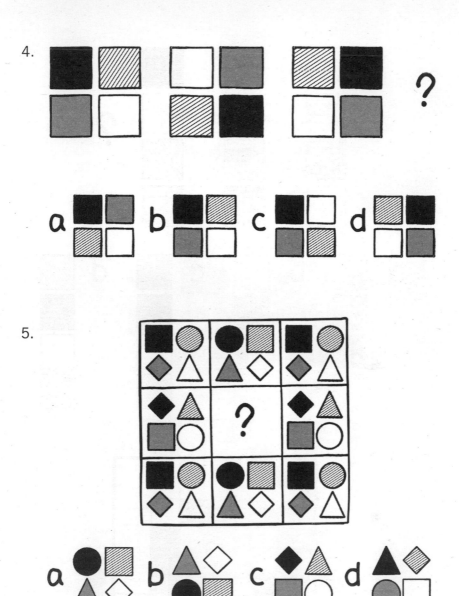

5.

6.

7.

8.

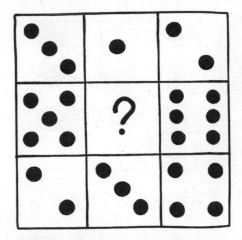

a 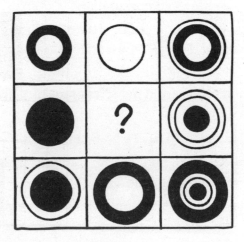 b c d

9.

a b c d

10.

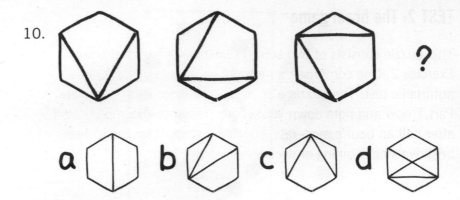

Answers

1. b

2. c (The bottom block is removed and added to the top of the next pile.)

3. a

4. b (The blocks are mirrored diagonally, then vertically, so next they need to be mirrored horizontally.)

5. b (The pattern mirrors adjacent shapes, and the colours alternate as dictated by the horizontal and vertical rows of shapes.)

6. c (The second object from the left is moved to the end each time.)

7. d ((2,4,6) – (3,6,9) – (4,8,12), so you need (5,10,15))

8. b (The numbers in the middle row are made by adding the numbers above and below them.)

9. c (The image in the left column is placed on top of and inside the image in the middle column to create the image in the right column.)

10. c (If the points of the hexagon are continually labelled 1,2,3,4,5,6,7,8, etc. clockwise from the top, the lines join points (1,3,5) – (2,4,6) – (3,5,7). So you're looking to join points (4,6,8).)

TEST 2: The brain game

This puzzle consists of two sets of questions – Exercise 1 and Exercise 2. One comprises a memory game and the other is an arithmetic test. The puzzle also consists of two parts. Complete Part 1 now and note down your score. You should complete Part 2 after half an hour's moderate exercise – good examples include brisk walking, running or cycling.

Part 1

Exercise 1: The memory matrix

Below is a series of grids. Some squares are coloured in black. Your task is to remember the pattern of the black squares. You only have three seconds to lock the pattern into your memory before you must cover the grid with a sheet of paper and recall the position of the black squares, marking them in the blank grid alongside. To ensure that you don't look at the next pattern, cover the rest of the questions with the sheet of paper and gradually move it down to reveal subsequent grids.

How many can you remember correctly (score 1 point for each grid)?

 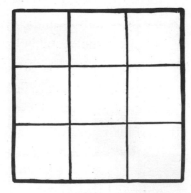

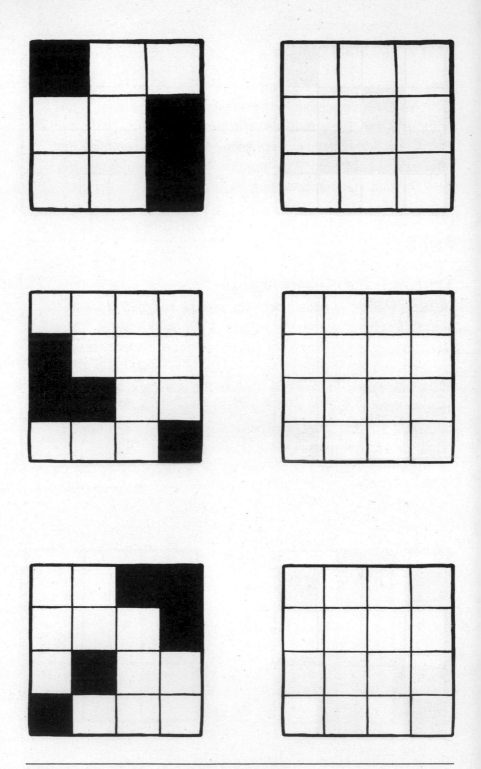

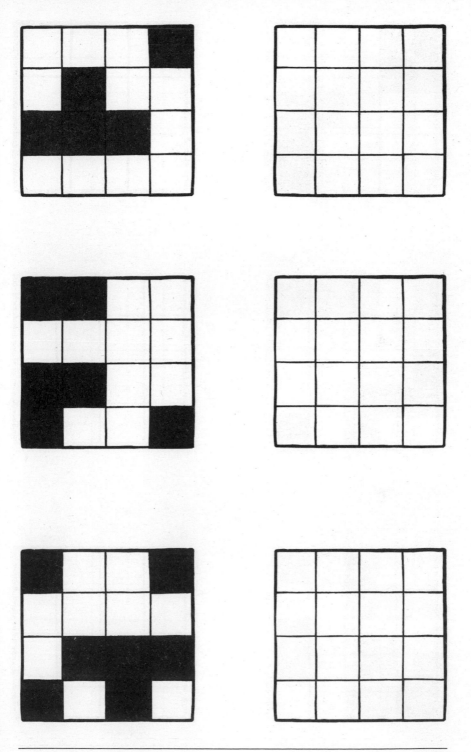

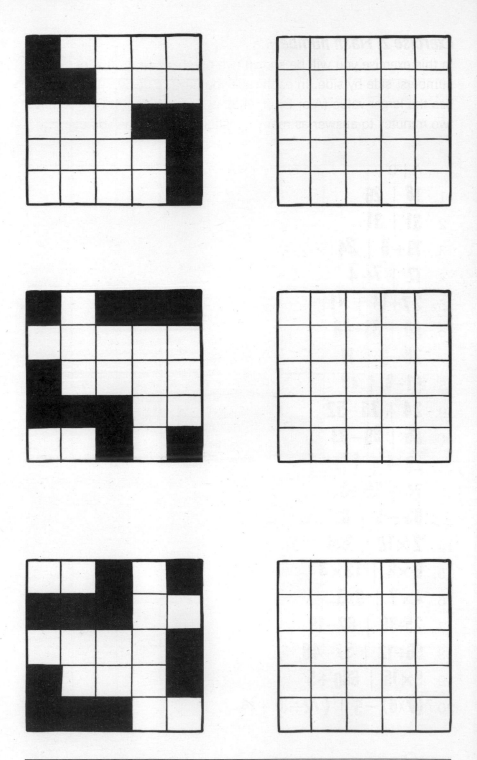

Exercise 2: Flash numbers

In this exercise you will be shown two numbers, or groups of numbers, side by side. In each case you have to decide which one has the larger value (A or B) or whether they are equal (C). You have two minutes to answer as many questions correctly as you can.

A | B

1. $18 \mid 26$
2. $31 \mid 31$
3. $19+6 \mid 24$
4. $12 \mid 7+4$
5. $27+14 \mid 41$
6. $80 \mid 33+48$
7. $28-9 \mid 18$
8. $51-8 \mid 43$
9. $24 \mid 75-52$
10. $65 \mid 99-33$
11. $20 \div 4 \mid 4$
12. $14 \mid 39 \div 3$
13. $63 \div 9 \mid 8$
14. $2 \times 10 \mid 3 \times 4$
15. $6 \times 6 \mid 12 \times 3$
16. $4 \times 7 \mid 9 \times 3$
17. $25+38 \mid 82-19$
18. $96 \div 12 \mid 57-48$
19. $9 \times 15 \mid 680 \div 5$
20. $(7 \times 6) - 5 \mid (72 \div 3) + 14$

Answers

1. B
2. C
3. A
4. A
5. C
6. B
7. A
8. C
9. A
10. B
11. A
12. A
13. B
14. A
15. C
16. A
17. C
18. B
19. B
20. B

Part 2 *(to be completed after 30 minutes of exercise)*

Exercise 1: The memory matrix

 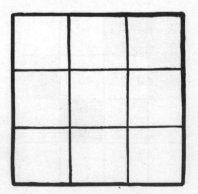

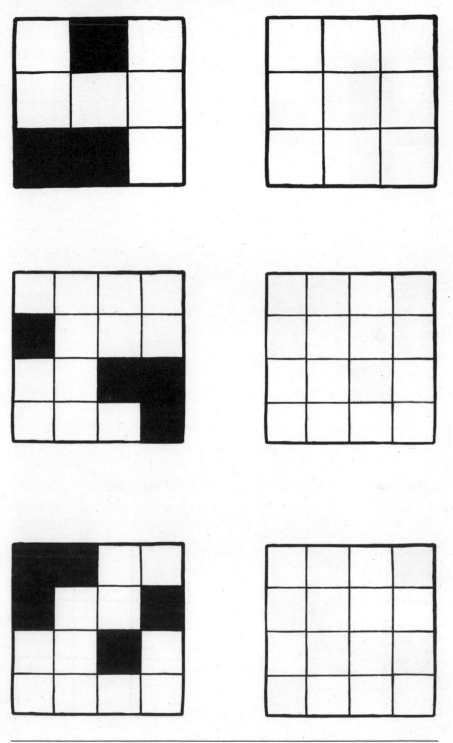

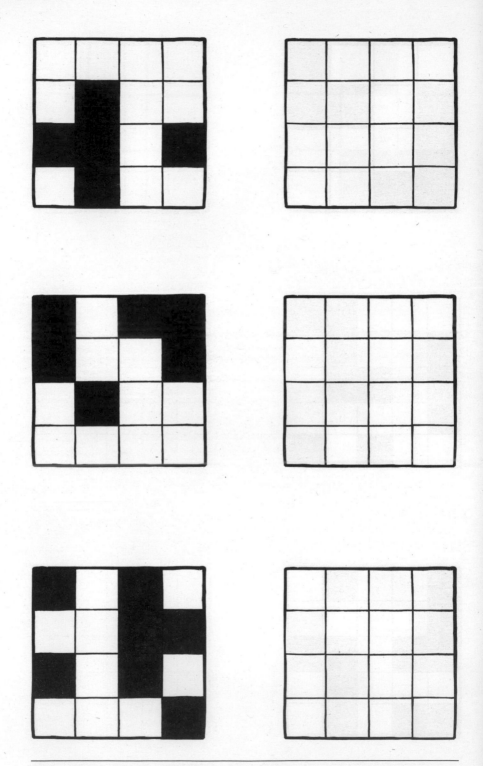

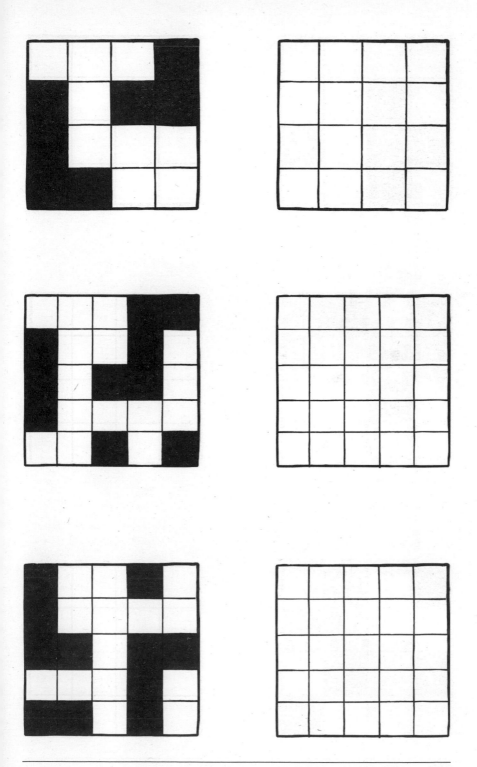

Exercise 2: Flash numbers

Which has the larger value (A or B) or are they equal (C)?

A | B

1. **25 | 17**
2. **41 | 42**
3. **21 | 9 + 17**
4. **8 + 7 | 14**
5. **45 | 29 + 16**
6. **52 + 19 | 72**
7. **15 − 6 | 8**
8. **84 | 93 − 8**
9. **63 − 49 | 14**
10. **37 | 77 − 38**
11. **28 ÷ 7 | 4**
12. **16 | 45 ÷ 3**
13. **54 ÷ 6 | 8**
14. **3 × 11 | 4 × 8**
15. **17 × 3 | 5 × 10**
16. **6 × 7 | 9 × 5**
17. **42 + 15 | 75 − 17**
18. **91 − 84 | 84 ÷ 12**
19. **512 ÷ 8 | 13 × 5**
20. **(105 ÷ 3) − 4 | (4 × 9) − 6**

Answers

1. A
2. B
3. A
4. A
5. C
6. B
7. A
8. B
9. C
10. B
11. C
12. A
13. A
14. A
15. A
16. B
17. B
18. C
19. B
20. A

Did you do better in Part 1 or Part 2? And in Exercise 1 or Exercise 2? Read on only if you've completed both tests.

TEST DEBRIEF

Unlike the astronaut selection-process questions in Part One of this book, it isn't your maths or memory skills that are being tested here. It is unimportant whether you did well in the tests or not. Instead it is the *difference* between your performances in the two sets of questions that space agencies are interested in.

These tests are very similar to those undertaken by participants in the Mars-500 project, a human experiment conducted by ESA, Russia and China in 2010 to simulate the psychological effects of a Mars mission right here on Earth (see page 224, where we will explore the experiment in more detail). The scientists running the project found that the crew did better in these types of questions after a period of exercise. Participants were asked the questions more frequently than you were in the exercise above, and over a longer period of time. So don't read too much into your results, but it does give you some idea of the sort of tasks that trainee Mars astronauts are put through.

Results from experiments like this will be fed into an exercise regime on board the first Mars mission, to keep the crew's cognitive functioning at the highest possible level. On lengthy space voyages, such as those required to reach Mars, it is important that all members of the team maintain their mental function. Mission Control wants to know ahead of time about the factors that are most likely to affect cognition.

The pattern-recognition exercise was very similar to the cognitive tests I had to do as part of my mission to the ISS. Each block of tests would last for 30 minutes and, in addition to pattern recognition, contained nine other types of puzzles. We were tested approximately ten times before we launched, to establish a baseline, and then every fortnight in orbit – typically half an hour before bedtime, when we were likely to feel most tired. Mission scientists were looking to see if our cognitive abilities changed over time and whether that was linked to our sleep patterns, workload, and so on. The tests also gave the crew immediate feedback on our performance, so we were able to see for ourselves if our mental functioning was at the expected level.

Tests like these are just one of the ways that work carried out on the ISS is helping us to prepare for a Mars mission. Mission planners can optimise our schedules to make sure we are working in the most efficient way. The same tests are conducted on participants taking part in missions at the Concordia Research Station in Antarctica, which also prepares candidates for long-duration space missions (see page 239). In future, the selection process may have to be adapted to test not only a candidate's cognitive ability, through the sorts of questions you experienced in Part One, but also whether those abilities are consistent over a considerable period of time.

TEST 3: Mood diary

A trip to Mars will see you far away from home and experiencing an unprecedented level of isolation. It is important to flag problems as soon as there are signs, rather than letting them develop into a bigger issue. This is just as true when it comes to psychology as it is concerning nuts-and-bolts equipment.

Your task over the next week is to keep a brief diary of your thoughts and feelings. Each day write a few lines about what you did, and another couple about your mood – whether you felt happy, sad, excited, nervous, and so on.

Every other day you should also record yourself on your phone or using a microphone, reading the following passage from an old Aesop's Fable known as 'The North Wind and the Sun':

> *The North Wind and the Sun were disputing which was the stronger, when a traveller came along wrapped in a warm cloak. They agreed that the one who first succeeded in making the traveller take his cloak off should be considered stronger than the other.*
>
> *Then the North Wind blew as hard as he could, but the more he blew, the more closely did the traveller fold his cloak around him, and at last the North Wind gave up the attempt. Then the Sun shined out warmly, and immediately the traveller took off his cloak.*
>
> *And so the North Wind was obliged to confess that the Sun was the stronger of the two.*

Later on in this part there will be more details about what to do with your diary entries and the recordings.

THE CHALLENGES OF MARS

The tests so far in Part Four may seem tricky or unconventional to you, but they are unusual by virtue of the fact that sending people to Mars will be the greatest undertaking that humans have ever attempted – a feat that requires rigorous and lateral thinking. Here are just some of the unique challenges presented to future astronauts by the Red Planet.

ISOLATION

Getting to the Moon may take three days in a spacecraft, but it's only one-third of a million kilometres away. Mars, on the other hand, is at least 50 million km distant – sometimes much further, due to the orbits of Earth and Mars around the Sun. That could mean a seven-month journey each way. It would take you longer than a standard mission to the ISS just to reach the Red Planet, let alone for you to explore it and return home.

You and your crew would experience a greater level of isolation than any human beings in history. Eventually the blue marble of Earth would fade into the distance, with the red orb of Mars yet to emerge from the inky darkness. If you get homesick on the ISS, at least you have the planet right there in the window below you. But on Mars, being considerably more cut off from the rest of humanity is likely to have a significant psychological toll.

Any selection tests recruiting astronauts to travel to Mars need to take this into account. Some space agencies in the future may even select a trained psychologist for missions, who will be able to help regulate the mood and emotions of the team. As we'll examine in a moment, space agencies around the world are already looking into strategies for coping with long-term isolation.

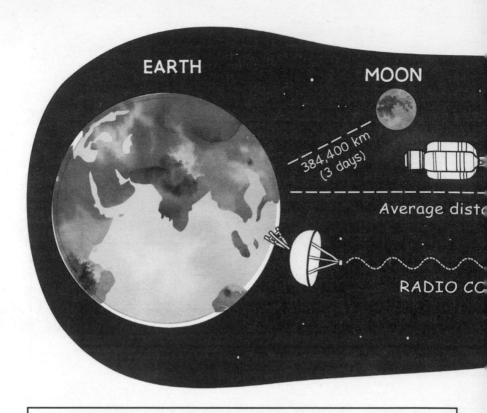

EARTH
MOON
384,400 km
(3 days)
Average dist
RADIO CC

AUTONOMY

When I was on the International Space Station we were under the constant, watchful gaze of Mission Control, which monitors everything astronauts do, and everything the ISS does, in minute detail. If anything goes wrong, Mission Control is right there, ready to help. That simply isn't an option on a Mars mission, because of the unavoidable communications delay.

Even though radio transmissions travel at the speed of light – a whopping 300,000 km/second – it still takes them 4–24 minutes to travel the Earth–Mars distance (depending on exactly where the planets are on their respective journeys around the Sun). Imagine encountering a problem halfway to Mars. It could take ten minutes for your Mayday to

▶

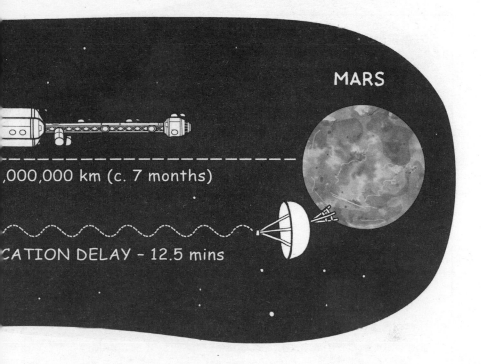

MARS

,000,000 km (c. 7 months)

CATION DELAY – 12.5 mins

reach Earth, and another ten minutes to get a reply. Space agencies have introduced artificial time delays into several training scenarios to learn more about the consequences of delayed communications. In fact, throughout my 12-day underwater NEEMO mission (see Photo Inset 2), a 50-second time delay was artificially imposed between Mission Control and the underwater habitat. It is possible that the ISS could be used in the future to train for deep-space missions by artificially increasing the levels of isolation and introducing a communication time delay.

This unavoidable consequence of the sheer vastness of space means that astronauts selected for a Mars mission will need to be highly skilled at decision-making and problem-solving on their own. Training in this area will also have to be adjusted to equip the crew with the right tools for the job.

RADIATION

Space is a dangerous place, not least because of the high-energy radiation flooding the solar system from the Sun and the wider universe. Thankfully, on the ground we are largely protected from this harmful energy by the atmosphere and by Earth's magnetic field. Even on the ISS we receive a considerable amount of shelter, since the space station's orbit lies within Earth's magnetic field. Parts of the ISS where astronauts spend larger amounts of time (for example, the crew quarters) are more heavily protected from radiation, and if the Sun expels a huge storm of protons (known as a 'coronal mass ejection'), the crew may have to take shelter there. Another form of dangerous radiation consists of galactic cosmic rays, or GCRs. These high-energy protons and atomic nuclei originate mainly from outside the solar system and can produce showers of secondary particles when they strike parts of the ISS.

When you are halfway to Mars, the relative safety of life within Earth's magnetic field is no longer on offer. A certain dose of radiation is sufficient to induce radiation sickness and can increase the risk of long-term conditions such as cancer and cataracts in your eyes. Space agencies around the world are still trying to develop lightweight ways of shielding astronauts who are journeying across the solar system. Water is an effective barrier, but its heaviness makes it far from an ideal choice. Other options include lining gaps in the walls of the spacecraft with dried, disinfected human waste – you would be making enough of it on a seven-month mission anyway, so you might as well use it. Whatever the solution, it is not inconceivable that Mars astronauts might be selected for their ability to tolerate higher levels of radiation.

LANDING

Landing on Mars is notoriously difficult. Looking back at our attempts to put robotic landers and rovers down on the surface, around half of the missions have ended in failure. A notable success saw NASA land the car-sized *Curiosity* rover there in 2012. Mission Controllers dubbed the final approach the 'Seven Minutes of Terror'.

This is where the effect of the communications delay really comes into sharp focus, particularly if it lasts longer than seven minutes. By the time Mission Control back on Earth receive the signal telling them you've hit the top of the Martian atmosphere, you'll already be on the ground. If something goes wrong during descent, there's nothing they can do to help.

Astronauts selected today for missions to the ISS know they are following a very well-trodden path. Yes, space travel is potentially dangerous, but a large number of the risks have now been mitigated, and trips to Low Earth Orbit are considered by many as being routine. A human mission to Mars would come with so many unknowns that the selection process would have to pick people willing to pay the ultimate price, if something went wrong.

SURVIVING: ISRU

You've landed successfully on Mars – a remarkable feat in itself. But landing on Mars is one thing, surviving on the Red Planet another entirely. Humans need a lot of sustenance, from food and water to oxygen and the removal of carbon dioxide. The more of these supplies you take with you from

Earth, the heavier your mission becomes. That not only means more expense, but landing a heavier spaceship on Mars becomes even more difficult. So the buzzword in space-science circles right now is 'ISRU' – or 'In-situ Resource Utilisation'. It means using as much of what's already on Mars as possible.

You could utilise Mars's abundant ice and potential water resources. Some architects have proposed outlandish designs of giant igloos, to be used as habitation structures. Scientists have proposed harnessing sunlight to prise open water's H_2O structure – two atoms of hydrogen married to one atom of oxygen – which would give you breathable air. The process of using electrolysis to separate water into hydrogen and oxygen is one that is already used extensively on the ISS. Equally, you could harvest the oxygen from Mars's thin carbon-dioxide (CO_2) atmosphere. Just such an experiment will fly out on the upcoming robotic Mars 2020 mission, to test out this latter theory about the carbon dioxide on the Red Planet.

The upshot of these inventive survival solutions is that the selection process for Mars astronauts will need to identify people who possess sound scientific and engineering credentials, which are capable of maintaining these crucial life-support systems with limited help from Earth.

Taking into account all these factors, a Mars crew will clearly need to consist of a range of highly skilled astronauts. Psychologists, medical doctors, engineers and scientists will all bring valuable skills as part of a diverse crew on a future mission. The way they are trained will also need to be adapted, to cope with the unique challenges that the Red Planet brings. Fortunately, preparations are already under way.

Here is one concept for a habitat on Mars, based on a 2016 NASA study. More recent concepts have also explored the use of 3-D printing using Martian regolith.

LANDER ON MARS

DOME INFLATED

COATED WITH WATER, WHICH FREEZES

PERMANENT HABITAT

MARS-500

Seven months journeying to Mars, at least a month-long mission on the surface and then an equally long return journey: the first people to walk on the Red Planet could be in space for around a year and a half – longer than anyone has so far spent beyond the Earth.

Understanding and investigating the challenges associated with such long-duration spaceflight was the goal of the Mars-500 project. Starting in June 2010, an all-male crew of six spent 520 days living in isolation inside a mocked-up spaceship at Star City in Moscow. That's the equivalent of a 245-day journey each way, plus 30 days on the surface of Mars to explore. They lived cheek-by-jowl in a restricted space with a total volume of just 550 cubic metres. Individual bedrooms were just 3 square metres each. Participants were cut off from communication with the outside world. Talking to a simulated Mission Control came with a 20-minute delay each way – typical of an interplanetary trip. They even simulated walking on the Martian surface, on arrival.

The experiment was a collaboration between ESA, China, the Russian space agency (Roscosmos) and the Russian Institute for Biomedical Problems (IBMP). The team included three Russians (Alexey Sitev, Sukhrob Kamolov and Alexander Smoleyevski), a Frenchman (Romain Charles), an Italian (Diego Urbina) and a Chinese citizen (Yue Wang).

Candidate selection

Would you want to spend more than 500 days isolated from the rest of the world? Particularly as you would see none of the real benefits and wonders of actually travelling into space. It seems plenty of people were willing to take on the challenge – the project received more than 300 applications.

While the process was a little less stringent than a real astronaut selection, there were still plenty of criteria that you had to measure up to, in order to be considered. Candidates had to:

- Be aged 20–50

- Be in good health and no taller than 185 cm

- Speak one of the working languages: English and Russian

- Have a background and work experience in medicine, biology, life-support-systems engineering, computer engineering, electronic engineering or mechanical engineering

- Be a national and resident of an ESA Member State participating in the European Programme for Life and Physical Sciences – that is: Austria, Belgium, Switzerland, the Czech Republic, Germany, Denmark, Spain, France, Greece, Italy, Ireland, Norway, the Netherlands, Sweden and Canada (the UK has since joined the programme, in 2012).

From those 300 applications, 28 candidates were selected for an initial telephone interview. That narrowed down the field to just ten highly qualified and experienced candidates, who then gathered at the European Astronaut Centre in Cologne, Germany in January 2010 to undergo an extensive medical examination, an in-depth psychological test and a personal interview with an expert panel, to determine qualities such as their motivation and their suitability for the task. In many ways it was similar to the real ESA astronaut selection process.

ESA's final four European candidates trained at the IBMP facility in Moscow for four months, before Romain Charles and Diego Urbina were chosen to join the four other prime crew members inside the specially designed isolation facility in Moscow.

What were the selectors looking for?

Do you think you have the mentality and skills for such an endeavour? The ideal candidates were robust, emotionally stable, motivated team workers who were open to other cultures and could deal with the slightly spartan lifestyle that you would associate with an actual space mission. Not only were individuals' personality traits very important, but there was also a need to combine different personalities and talents, in order to create the optimal group for such an extensive exercise.

TIMELINE OF THE MARS-500 MISSION

On pages 227–8 is a timeline of the simulated mission that was conducted during the Mars-500 experiment. This schedule shows one prediction of what a mission to Mars may look like.

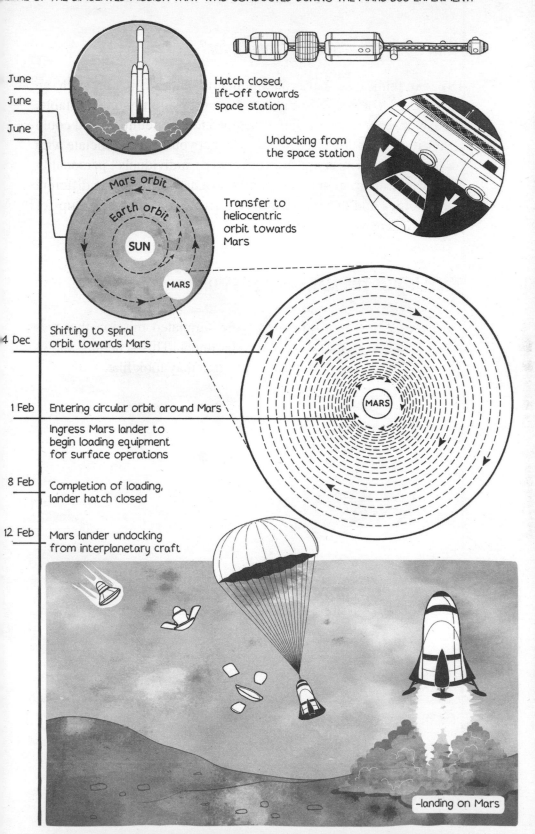

June — Hatch closed, lift-off towards space station

June — Undocking from the space station

June — Transfer to heliocentric orbit towards Mars

Mars orbit
Earth orbit
SUN
MARS

4 Dec — Shifting to spiral orbit towards Mars

1 Feb — Entering circular orbit around Mars

Ingress Mars lander to begin loading equipment for surface operations

8 Feb — Completion of loading, lander hatch closed

12 Feb — Mars lander undocking from interplanetary craft

–landing on Mars

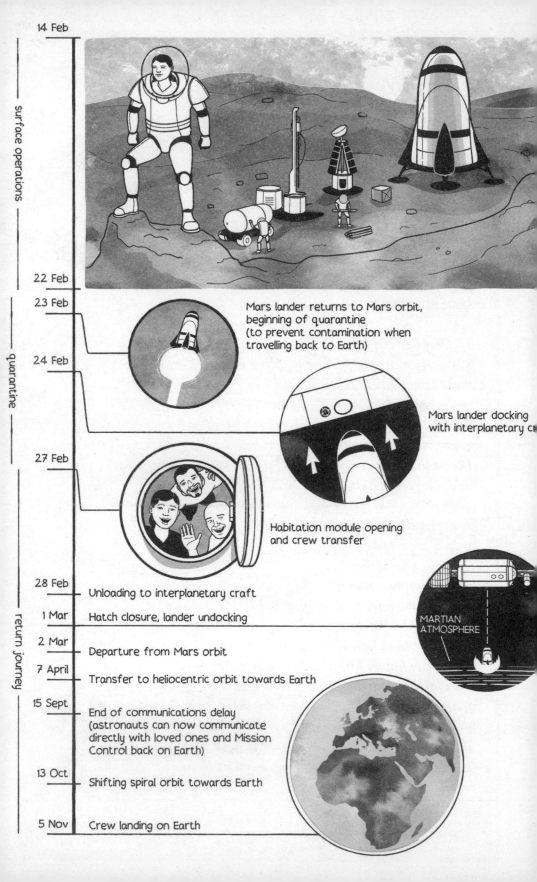

surface operations

14 Feb

22 Feb

quarantine

23 Feb — Mars lander returns to Mars orbit, beginning of quarantine (to prevent contamination when travelling back to Earth)

24 Feb

Mars lander docking with interplanetary c▮

27 Feb — Habitation module opening and crew transfer

return journey

28 Feb — Unloading to interplanetary craft

1 Mar — Hatch closure, lander undocking

2 Mar — Departure from Mars orbit

7 April — Transfer to heliocentric orbit towards Earth

15 Sept — End of communications delay (astronauts can now communicate directly with loved ones and Mission Control back on Earth)

13 Oct — Shifting spiral orbit towards Earth

5 Nov — Crew landing on Earth

MARTIAN ATMOSPHERE

Mars-500 diary

The Mars-500 'astronauts' had to keep regular diaries as part of the experiment. These provided a valuable record of daily events, as well as a picture of the mental states of the human subjects. In the diary entry below, written on 13 October 2011 – a date just shy of a whopping 520 days in isolation – the Italian crew member Diego Urbina gets philosophical. From describing the birthday meal of Sasha, one of his crewmates, to musing on the importance of the arts on space missions, it's a touching personal account. Can you imagine yourself in Diego's shoes? Would you be so controlled in the same situation?

13 October 2011
It's early morning in the spaceship modules. Of course, there is no way to tell it is morning except by the sounds of the other guys' blood pressure machines making that noise: 'Brrrrrrr'. The pressure is read aloud by a female voice in the machine, telling you how healthy you are. Blood pressure can go up with stress, salt consumption and many other kinds of things. No stress today: it's almost day 500 out of 520, and if your blood pressure or heart rate is exceptionally high, it's surely not stress: it would have to be your heart keen to land on Earth.

Today is Sasha's birthday. He always stays some more minutes in his cabin (or 'kayutka') before breakfast, so we are sure we have time to prepare the typical birthday treat on the table. We do so, and Sasha comes out of his room with his blond hair a bit messed up but with a huge smile as we say happy birthday in at least four languages.

Gifts for him include a T-shirt and some of his favorite food that he could have anyway by going to the storage module, but he quickly gets the joke: everyone has some food they are keen on, sometimes comically so. His favorite food items are Russian Shi soup and chocolate bars.

Other more elaborate gifts include a big poster with
Chinese calligraphy expertly made by our crewmate Wang.
He has practiced this ever since we started the simulation.
I am fascinated by this form of art because it seems so
ancient and full of history. I think he could already make a
living out of those beautiful posters, if of course he wasn't
too busy simulating trips to Mars with us or preparing
Chinese astronauts to go to outer space!

My gift is a poster with a pencil drawing I made of
Sasha when we were on the simulated Martian surface.
Although I'm not by any stretch of the imagination a
real artist, I think this drawing came out pretty cool, so I
thought Sasha would like it.

Art is a nice thing to have here because it gives some
personal satisfaction and makes you feel 'more human'.
Romain is a really good guitar player – we had our band

during the whole trip and had a lot of great moments with that. Fortunately for mankind, though, we were playing in space, where 'no one can hear you sing'.

Sukhrob has some paintings done by his daughter on his wall, they are pretty and show that she really feels like her dad is coming from Mars or some distant place, a lot of contrast with Doc Kamolov's way more realistic and practical ways! But I sure love those drawings his kids left in the ship.

I think you do need artists (real ones) in space. I'm not joking. Yeah, sure you need us engineers, doctors or scientists or you probably would be grounded, but I think ultimately it will be even more important to have painters and poets on Mars.

I remember when I was in school and I saw the landing of the ESA Huygens probe on the surface of Titan (one of Saturn's moons). The images were just baffling. Just a few seconds of video where every second that passed you discovered more and more of an alien surface, something nobody has seen before was being revealed with every passing moment. It was pure beauty to me. I am sure it was quite an adventure for the probe to go there and get all that data so we could better understand this mysterious moon. But . . . did this marvelous human-built probe appreciate what it was seeing? Did it feel chills when parachuting into such a fascinating place? Hard to say it was so.

One of the particularities of humans is their ability to translate what we see, hear, experience, into a 'media format' with a soul: into art.

What would happen if Leonardo were sent on a mission to Mars, and could see a sunrise from the top of the highest mountain in the solar system, Olympus Mons? What would happen if Michelangelo were transported to see the Sun traversing the rings of Saturn, arguably the *prettiest show of the solar system? What if they had their oils and a canvas?*

Sasha continues opening his gifts, somebody makes a comment on how incredibly close his birthday is to the end of the mission (the last of anyone inside the mission) and how absurdly far away this day seemed at the beginning, when we were having our first birthdays. Time doesn't fly, but everything has an end.

Now that it is coming to an end, I am still convinced that this was not a journey into the cosmos, but a journey to know ourselves and our minds, to realise how important respect and communication are in order to achieve a functional crew, how fundamental are the links to the real world, thin and fragile as they may be in this situation.

There are few other places where your humanity becomes as evident and all the connections that you have with others, within the crew and outside, become as 'naked to the eye', and are so in need of good care in order to maintain them. We somehow ended up feeling a little bit more human than normal, by having been taken 'away from humanity'.

I think it is an interesting experience to live through. Not necessarily one you'd enjoy 100% but you can make it livable and happy if you have the right mindset. The trick can be in trying to make this time 'better' (more accomplished, more well-utilised, etc.) than life outside.

It can also offer a good challenge if you look for ways to use the environment to your advantage. Forget about the things you don't have and squeeze all the juice out of the things that you DO have in isolation, optimising them! That was my approach and I enjoyed many moments by having overcome the difficulties that were in front of me, helping others overcome theirs, and letting others help me overcome mine. Satisfactions like these are what keep you going, they are sneaky little things, but they are all over the place waiting to be caught.

Sasha has breakfast and goes to his kayutka, waiting for messages from the ground during one birthday he will surely not forget.

Diego

In focus: Romain Charles

Frenchman Romain Charles was the other crewmate selected alongside Diego as one of the two European participants of the Mars-500 project. Here he tells us, after the mission, what it was like to go through the selection process and to spend 520 days confined inside the experimental facility in Russia.

Q *What surprised you most about the selection process?*

A In the 11 candidates there were two clear leaders. They were very good at it and all of us expected one of them to be appointed the commander of the mission, which was okay because they were great people. In the end, neither of them was chosen. Instead the commander was somebody who is full of common sense. He would lead by example and make things easy to do. It worked for us – having such a commander led to the success of the mission.

I also learnt after the mission that selectors were looking for optimistic, creative people. It made a lot of sense in hindsight, because you have a lot of free time and not so many things to fill it. Creative people can find interesting ways to pass the time. Optimistic people often look for solutions to problems, try to move forward and tend to look on the bright side.

Q *What sort of activities formed part of your team training?*

A Imagine a tube with a big fan below it and a platform in the centre with an inflatable tyre. We had to operate this device as a team of two. If you were standing in the correct position you would go up, or if you leant one way the platform would go down. The idea was that we had to work together to move the platform. Training like this helped us to get to know each other. I feel we could have had more team training, though. We also bonded with some people who didn't end up being part of the crew and we had to say goodbye.

Q *Do you have fond memories of the experience?*

A Like in the outside world, you have ups and downs, but overall I would say it was positive. Emotions were a bit more amplified towards the end of the confinement. Feeling good felt great; feeling bad felt worse. It was a success, though – we showed that a crew can endure the confinement associated with a mission to Mars.

Q *How did you deal with the extreme isolation, and what did you miss the most?*

A When I entered the experiment I made a conscious deal with myself that anything I found I couldn't do, I would do in a year and a half. So any cravings I had, or things I missed, I would just think: I will eat that, or do that, when I get out. After a couple of days the craving would go away.

I remember really wanting to feel water all around me – to jump into the sea or a swimming pool. This lasted for a couple of days. Another big one was linked to sound. I kept imagining and reimagining the sound of the wind blowing through the fields of my uncle's farm. We would holiday there during harvest season when I was a child. The sound the wind makes through the ripe wheat was often the only thing you could hear. I missed this and regretted not taking a recording of it in with me.

Q *How was the food?*

A Good. It got a little more boring as we lost variety over the course of the mission, but in itself it was okay. It was only after, when I tasted fresh food again, that I realised how good food is when it's fresh. So I didn't really miss food in particular – it just made me enjoy things more when I got out.

Q *How do you deal with conflict, when you can't just get up and walk away?*

A Firstly, we were prepared in training for this to happen. The psychologists told us that it was normal and likely to occur. The truth is that we didn't have conflict during the mission. There were tensions, but it never escalated into a conflict, because we were able to communicate early enough. We were open enough to talk to each other without waiting, and willing to accept remarks from crewmates. That was sufficient to defuse any potential conflict.

Q *What was the greatest source of tension?*

A It was mostly due to cultural differences, particularly around mealtimes. For one of the Russian crew members, it was traditional to be silent while eating, with conversation only after the meal during tea or coffee. For me, a meal is a very important moment where you talk and you share with the group. He got upset by me asking him questions while we were eating. At first I was taken aback, because I didn't expect such a reaction and the meal ended a bit awkwardly. But within ten minutes he came to me and explained why he'd snapped at me, saying that mealtimes for him are different. Once we understood the difference, we could reach a compromise and it was fine. I would ask him questions at the end of the meal, and it worked well for the rest of the mission.

TEST 4: Coloured Trails

One of the experiments conducted by the Mars-500 crew involved playing a three-player negotiation game called Coloured Trails. The results of their game-play are being fed into what is known as a Mission Execution Crew Assistant (MECA). MECA is a bit like an advanced version of AI-based assistants such as Siri and Alexa that are currently on the market. It is hoped that the AI interface will help humans make the right decisions and deal with unexpected, complex and potentially hazardous situations during long-duration missions like those to Mars. The goal of the project is to encourage human and machine groups to act in an autonomous, but cooperative way. During the playing of Coloured Trails the human players were sometimes given advice by the system.

The Coloured Trails board consists of a grid of coloured squares (red, yellow, green, blue and orange – marked by their initial letters in the grid opposite). All three players begin on the square marked 'Start' and must try and reach the square marked with their name – i.e. Player 1 must aim for the square marked 'Player 1', and so on. Moves can only be made one square at a time and by moving horizontally or vertically (diagonal moves are not allowed). Each player is in possession of four coloured playing chips, randomly selected from five of each colour. A player may only move to an adjoining square if they hold a chip of the corresponding colour – e.g. you can move into an orange square if you hold an orange piece.

Before any moves can be made, there is a negotiation round. If you don't hold the chips you need, you can negotiate with the other players to obtain them. In each round, two of the players take on the role of 'proposers' while the third player is a 'responder'. The two proposers are each able to suggest a trade to the responder (but they don't have to). The responder is able to accept one of the proposals, but not both, but does not have to accept either proposal. If the responder accepts a proposal, the appropriate counters are traded. Every player then makes one move, before another negotiation round begins. The role of responder moves to the next player in the sequence, so that everyone has an equal turn at it.

Each game lasts 15 minutes. If a player reaches their target before the time is up, they receive 50 points. Everyone else is docked 25 points for every square they are away from their target. If no one reaches the target, then everyone is docked 25 points for every square they are from their target.

To try this at home with a team of three players, decide beforehand how many games you are going to play. Use the grid below. The winner is the one with the most points at the end.

PLAYER 1 R	G	R	B	B
R	Y	O	Y	R
G	R	START R	Y	O
R	O	O	B	PLAYER 3 B
PLAYER 2 Y	G	R	Y	Y

PLAYER 1 (O) (B) (B) (G)

PLAYER 2 (B) (B) (Y) (Y)

PLAYER 3 (O) (G) (R) (R)

Mars-500: Conclusion

Based on Mars-500, do you think you could endure a mission to Mars? As you will have realised, the unique challenges of a mission to the Red Planet will demand extraordinary psychological robustness from our future astronauts, in addition to the many other skills that have been covered in this book. During the final weeks of my mission to the ISS, a leaking Progress cargo spacecraft meant that the space station's external window covers had to be kept closed to protect them. While this was nothing in comparison to a trip to Mars, it was interesting to witness the sudden change in environment. Sunlight no longer flooded into the modules during the daytime, it became harder to keep track of time, and there was no quick psychological boost from a trip to the Cupola window to marvel at the view or take some photographs. The artificial environment inside the sterile laboratories became more oppressive, and for the first time in nearly six months in space I began to feel slightly confined. It was enough to make me appreciate that the research we are doing into the psychology of deep-space missions will play a vital role in ensuring the well-being of those first human explorers to visit the Red Planet.

Another fascinating long-duration human experiment that is shedding new light on what it will take for astronauts to live on other planets consists of the isolation studies conducted in one of the most remote locations on planet Earth – Antarctica.

CONCORDIA RESEARCH STATION

Few places on Earth are as remote as the Antarctic. That makes the Concordia Research Station, tucked away on an ice plateau 1000 kilometres from the open ocean, a great place to work on isolation studies in preparation for long-duration human spaceflight. Medical doctor Beth Healey spent a year living and working there, on behalf of the European Space Agency. Here she tells us what life was like, being so cut off, and how that can help us prepare for a journey to Mars in the not-too-distant future.

Q *How were you picked to go to Antarctica?*

A Just like the astronauts, I had to go through a lot of medical and psychological testing after the initial application. I also had to do the famous Rorschach test as part of a deeper interview with a psychologist. Before the crew travelled to Antarctica we went through Human Behaviour and Performance (HBP) training at the European Astronaut Centre – again just like the real astronauts.

Q *What were you doing when you applied?*

A I'd just finished my junior-doctor training, working in London, but I'd already been doing some medical work in extreme environments. I worked in Greenland for three seasons, and had been part of the medical support team for both the Black Ice Race in Siberia and the North Pole marathon. So I was used to working as part of a team in these challenging settings.

Q *What makes Concordia a great stand-in for long-duration human spaceflight?*

A The temperature, for starters – it never really gets above –70°C. There are also 105 days of the Antarctic winter when the

Sun never rises. These two factors made us inaccessible, in terms of evacuations for that period. No planes were able to fly to come and get us. That's true isolation. Currently, on the International Space Station, an astronaut can be evacuated in half a day if they have a medical emergency. On the way to Mars that wouldn't be possible, which is why we're using places like Concordia to look at the psychological effects that isolation has on a crew.

Q *What was the hardest part of the experience for you?*

A The darkness was a real struggle. It completely threw my body's natural sleep–wake rhythm and I ran into a lot of trouble. That was the same for most of the crew. It's really hard to artificially re-create the same light as the Sun. You don't wake up or sleep properly – you go into semi-hibernation. The Sun is such a familiar feature that, when you lose it, you feel really disconnected from life back home. When it came back, the whole crew had a huge morale boost. We had so much more energy almost instantaneously, even though it only popped up above the horizon for a few minutes to start with.

Q *What was the best bit of your time there?*

A Also the darkness, weirdly. It meant that we saw stunning auroras (Southern Lights) and beautiful night skies. It's like nothing else on Earth – you just walk out of the door at breakfast time or lunchtime and you see the full Milky Way arching overhead. Also, the adventure of being in Antarctica over winter was something I'd grown up thinking about. To feel like I was contributing to the future of human spaceflight was also something quite special.

Q *How was the team dynamic, given that you were living in such close quarters?*

A It was really interesting, because you don't choose who you go down there with. I was most concerned beforehand with being lonely, but that turned out not to be an issue. It's kind of the opposite – you can't get away from people. I also thought there'd be lots of open conflict, people arguing with each other and everything being quite dramatic. In fact it was the opposite. Nobody wanted to be seen as the bad guy, so you have lots of undercurrent behaviour. A good example is that somebody who didn't like me used to hide all of my stuff. I never quite knew who it was, but several months of it is like a psychological minefield. That was pretty classic of the sort of behaviour you'd see, rather than big open fights. People found other ways to get at those they weren't so fond of. Everyone got very sneaky.

Q *Why do you think it was like that?*

A Everyone's behaviour was so heavily scrutinised by the rest of the group that we became very conscious of how our actions would be perceived by others. As a young woman, I found that some people were actively unkind to me in front of the group, so that they didn't get teased for fancying me. People didn't want to be seen to be overly nice to someone. This also applied to situations like sucking up to the station leader. It was much more pronounced than you get in normal life. Living in isolation magnifies everything a lot more.

Q *Based on your experience, what advice would you give someone travelling to Mars in the future?*

A Be clear about your motivation and why you want to do it. You can return to this, if you're struggling at any point. I was really passionate about the science we were doing, so that became a big focus for me. I could be clear about why I was there. If you're doing it for the money or the kudos, that is harder to fall back on – it will wear off quite quickly.

The most important thing, however, is to have at least one really good friend. As long as you have one person that you really get on with, I think you can handle anything. The people who struggled at Concordia were those who semi got on with everyone, but didn't have someone they could just tell everything to.

TEST 3: Mood diary revisited

On page 216 I asked you to keep a brief diary of your daily activities and feelings for a week, along with recordings of yourself reading a brief extract from the Aesop Fable 'The North Wind and the Sun'. Read the following text only if you've completed those tasks. If not, you can move on to Test 5 while you complete the mood diary and the recordings.

Scientists have long known that speech is a great indicator of changes to a person's psychological state. On a mission to Mars it is particularly important to pick up early warning signs if a crew member is experiencing a period of low mood. An experiment known as Psychological Status Monitoring by **Co**mputerised **A**nalysis of **La**nguage phenomena (COALA) has been conducted on the Concordia Research Station in Antarctica in order to learn about the link between speech and state of mind during periods of extreme isolation. Participants completed video diaries, along with clips of them reciting the same passage from 'The North Wind and the Sun'.

These recordings were then compared to two databases of other people also reading the same fable. One database came from people with Seasonal Affective Disorder (SAD) – a type of depression sometimes called 'winter depression'. The other database contained readings from healthy people who had scored low on a questionnaire designed to diagnose depressive illnesses. The researchers found that factors such as speech rate, length of pauses and intensity show significant changes in the speech of depressed people.

Can you spot any link between your thoughts and feelings in the diary and the different ways you read the passage? This level of self-awareness is a good skill to have, for future Mars astronauts. Remember, though, that the diary-keeping exercise you did is only a small sample and should never be used to diagnose any medical condition.

When it comes to a Mars mission, research like this carried out at Concordia could be the basis for a system that triggers an alert, if the speech of one of the crew members changes in such a way that it indicates an increased risk of low mood.

TEST 5: Rorschach test – what do you see?

In 1921 the German psychologist Hermann Rorschach invented a psychological test based on a number of inkblot images. During the astronaut selection process we were given the same test during the hour-long one-on-one psychological interview, so that the selectors could examine our personality characteristics and emotional functioning. Beth Healey was also given the same test when she was selected to travel to Concordia. So there is a good chance that it could form part of any future selection process, too.

The Rorschach test is conducted by professionals under specific conditions, which cannot be replicated fully in this book. But have a look at the following images – what do you see? It is a very subjective experiment, open to a wide amount of interpretation. Is your answer one of the most common, or do you see things differently from most?

Most common answers

1. bat, butterfly, moth
2. animal hide, skin, rug
3. human heads or faces (especially women and/or children)

The Rorschach test is famously vague – you can see just about anything you can imagine. You can also worry too much about what you think the selectors want you to see. In our case, during the 2008 selection process, there was no right or wrong answer that the selectors were looking for. It's better just to be honest.

Back in the early days of spaceflight, NASA astronaut candidates for the Mercury programme to put the first American in space were given the same tests. Candidates Alan Shepard and Pete Conrad openly discussed with each other what the instructors might be looking for. Shepard, who would go on to become the first American in space, convinced Conrad that virility was the order of business. So Conrad gave sexual answers when shown every card. His disdain for the tests was underlined when he was shown a blank card and quipped that he couldn't interpret it because the psychologist was holding it upside-down. Unfortunately for Conrad, the medical panel did not share his humour and he didn't make the grade that time. He was, however, later selected for the Apollo programme and became the third astronaut to walk on the Moon, when he commanded the Apollo 12 mission in November 1969.

CONCLUSION: ASTRONAUT GRADUATION

As you will have discovered during the course of this book, getting hired as a space explorer takes skill, perseverance and hard work. In the real world, it also takes patience and a fair share of luck. To maximise your chances, here are some final words of advice if – as Yuri Gagarin put it – you choose to follow the 'Road to the Stars'.

This is never a first career. Astronaut selectors want to see evidence that you've picked up a considerable number of skills across a wide variety of disciplines, so don't expect to be selected until you're in your thirties. This is why it is so important to find out what motivates you early in life and to follow that path. If you are truly passionate about something, then you are far more likely to be successful in that first career choice. You will need a relevant degree and years of work experience in a technical area. The intensive selection process then puts your teamwork, problem-solving and communication skills to the test. Get through those tests and you'll face an array of medical examinations. Many of the qualities that the space agencies are looking for in those early stages are determined by your personality and are characteristics that cannot be learnt. If you can't control something, then try not to worry about it. Instead, enjoy it as a rare chance to see how you measure up to one of the most scrutinising selection processes on the planet.

Few jobs are so all-consuming, so if you're successful, expect to hand over your life – and your family's lives – to the cause. You may well have to move countries, learn new languages and spend years training. As an ambassador for space exploration, there will be considerable attention from the media and you will have to be comfortable with frequent public engagements – something your previous career may not have prepared you for.

Yet your reward is an experience of unrivalled adventure and wonder. Memories of free-floating through space and jaw-dropping views of Earth will stay with you for ever. Doing a spacewalk is like nothing you'll experience ever again. You will become part of an incredible international team of people who are pushing the envelope, laying the foundations of a permanent human presence in the universe at large.

As we have examined in this book, what space agencies are looking for from their astronauts has changed, depending on the mission. Gone are the days when an astronaut was an alpha-male fighter pilot full of machismo. Much has been made of the early space travellers and how they were made of the 'right stuff'. The average Apollo astronaut was male, 32 years old, a former jet fighter pilot, owned a Corvette and was married with two children and a dog. Modern space travel is different. Space agencies are no longer after fearless individuals for short, risky flights. Instead they are looking for calm team players, who are capable of living and working together in an isolated environment for months at a time. Factors such as gender and race are irrelevant. If you're good enough, that's it. Ultimately, today's astronauts are testing out the techniques and technology that will allow us to send people to Mars later this century – it is a privilege to be part of such an endeavour.

You might think you've either got what it takes or you haven't. While that may be true for some personal traits, it is possible to learn some of the skills required. I did. Being an astronaut means having a high level of self-awareness, and knowing your strengths and weaknesses. Try to have a growth mindset – see failure as an opportunity to learn and develop. Keep at it. Determination in the face of adversity is a wonderful skill to have, and one that selectors value. Astronauts don't cross their fingers and hope for the best. That's just not how we deal with risk. Be meticulous in everything you do.

To be part of that first team of explorers who set foot on Mars would be the most exhilarating feeling imaginable, extending human presence in our Solar System and paving the way for a future colony. Good luck – and aim high!

ACKNOWLEDGEMENTS

First and foremost, this book would not have been possible without the enormous dedication, expertise and generosity shown by everyone at ESA who has collaborated on the project. Special thanks to Rosita Suenson for all her efforts in making the book happen, for her ideas, and for assembling a super team of contributors. To all of those who were interviewed about the selection process, your fascinating insights have shaped this book, and will shape the future of human spaceflight. Particular thanks to Loredana Bessone, Romain Charles, Antonio Fortunato, Beth Healey, Ruediger Seine, Gerhard Thiele, Diego Urbina, Guillaume Weerts and Dr Iya Whiteley (UCL). To Julien Harrod, thank you for supporting the content of the book and for all your suggestions. To Carl Walker, thank you for all your helpful edits.

To Colin Stuart, for your brilliant and tireless work helping to write and research the book. To Ed Grace, for the fantastic illustrations that subtly communicate some very complex ideas. To Roger Walker, for designing everything up so beautifully. To Suzanne High, for your wizardry on the logic puzzles. To Dr Katie Steckles, for meticulously checking many of the most fiendish puzzles in the book.

Finally, I would like to thank the wonderful team at Cornerstone, Penguin Random House for their hard work and dedication to the project: Ben Brusey, Joanna Taylor, Linda Hodgson, Jason Smith, Charlotte Bush, Rebecca Ikin, Sarah Ridley, Sarah Harwood, Alice Spencer, Dan Smiley, Mat Watterson, Claire Simmonds, Kelly Webster, Cara Conquest, Pippa Wright, Catherine Turner, Khan Lawrence, Selina Walker and Susan Sandon.

PHOTOGRAPH CREDITS

Inset 1
Page 1, Survival Training: *All photographs:* © ESA – V. Crobu
Page 2, Survival Training (continued): *All photographs:* © ESA – V. Crobu
Page 3, Survival Training (continued): *All photographs:* © GCTC / ESA
Page 4, Survival Training (continued): *All photographs:* © GCTC / ESA
Page 5, CAVES: *Top photograph:* © ESA – R. DeLuca. *Bottom photograph:*
© ESA / NASA
Page 6, CAVES (continued): *All photographs:* © ESA – V. Crobu
Page 7, CAVES (continued): *Photograph:* © ESA – V. Crobu
Page 8, CAVES (continued): *Top photograph:* ESA – V. Crobu.
Bottom photograph: © ESA / NASA

Inset 2
Page 1, Neutral Buoyancy Training: *Top photograph:* © ESA – E.T. Blink.
Middle photograph: © ESA – H. Rueb. *Bottom photograph:* © ESA –
H. Rueb
Page 2, Neutral Buoyancy Training (continued): *All photographs:* © ESA –
S. Corvaja
Page 3, NEEMO: *All photographs:* © ESA – Hervé Stevenin
Page 4, NEEMO (continued): *All photographs:* © ESA – Hervé Stevenin
Page 5, Centrifuge Training: *All photographs:* © GCTC / ESA
Page 6, Centrifuge Training (continued): *All photographs:* © ESA – S. Corvaja
Page 7, Zero-g Training: *Top photograph:* © ESA – A. Le Floc'h
Middle photograph: © Novespace / ESA. *Bottom photograph:* © ESA –
A. Le Floc'h
Page 8, Zero-g Training (continued): *Top photograph:* © ESA – A. Le Floc'h
Middle photograph: © ESA / NASA. *Bottom photograph:* © ESA / NASA

Page 117, facial expressions: © Photo Researchers/SCIENCE PHOTO
LIBRARY

Endpapers
Earth: © ESA / NASA. Mars: © Phil James (Univ. Toledo), Todd Clancy (Space
Science Inst., Boulder, CO), Steve Lee (Univ. Colorado), and NASA/ESA

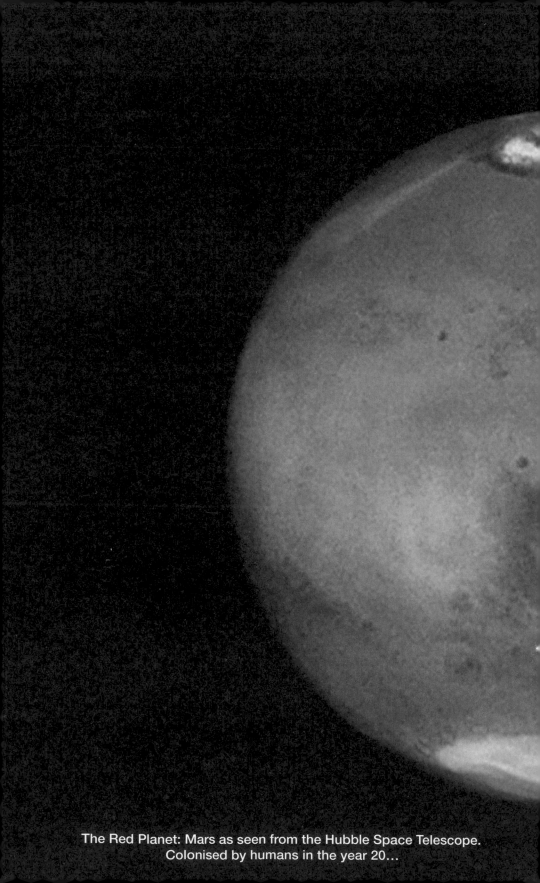

The Red Planet: Mars as seen from the Hubble Space Telescope.
Colonised by humans in the year 20...